Terry J. Williams is a full-time writer and illustrator. She worked as a tutor in publishing studies at Stirling University before spending ten years in a crofting community on the isle of Skye. She has written and illustrated feature articles for several magazines, including the *Scots Magazine* and the North American quarterly *Scottish Life*, with whom she continues to work as a regular contributor. *Walking With Cattle* is her second full-length work. Her first book, *Ten Out Of Ten*, was published in 2010 to commemorate the tenth anniversary of Sgoil Chiùil na Gàidhealtachd (the National Centre of Excellence in Traditional Music) in Plockton. She lives in the Black Isle near Inverness.

# WALKING WITH CATTLE

## IN SEARCH OF THE LAST DROVERS OF UIST

TERRY J. WILLIAMS

BIRLINN

First published in 2017 by
Birlinn Limited
West Newington House
10 Newington Road
Edinburgh
EH9 1QS

*www.birlinn.co.uk*



ISBN: 978 1 78027 488 1

British Library Cataloguing-in-Publication Data
A catalogue record for this book is available from the British Library

Typeset by Edderston Book Design, Peebles
Printed and bound by Grafica Veneta, Italy

'The brown sails of the cattle boats have gone from the Minch.'

A. R. B. Haldane

# *Contents*

# *List of Illustrations*

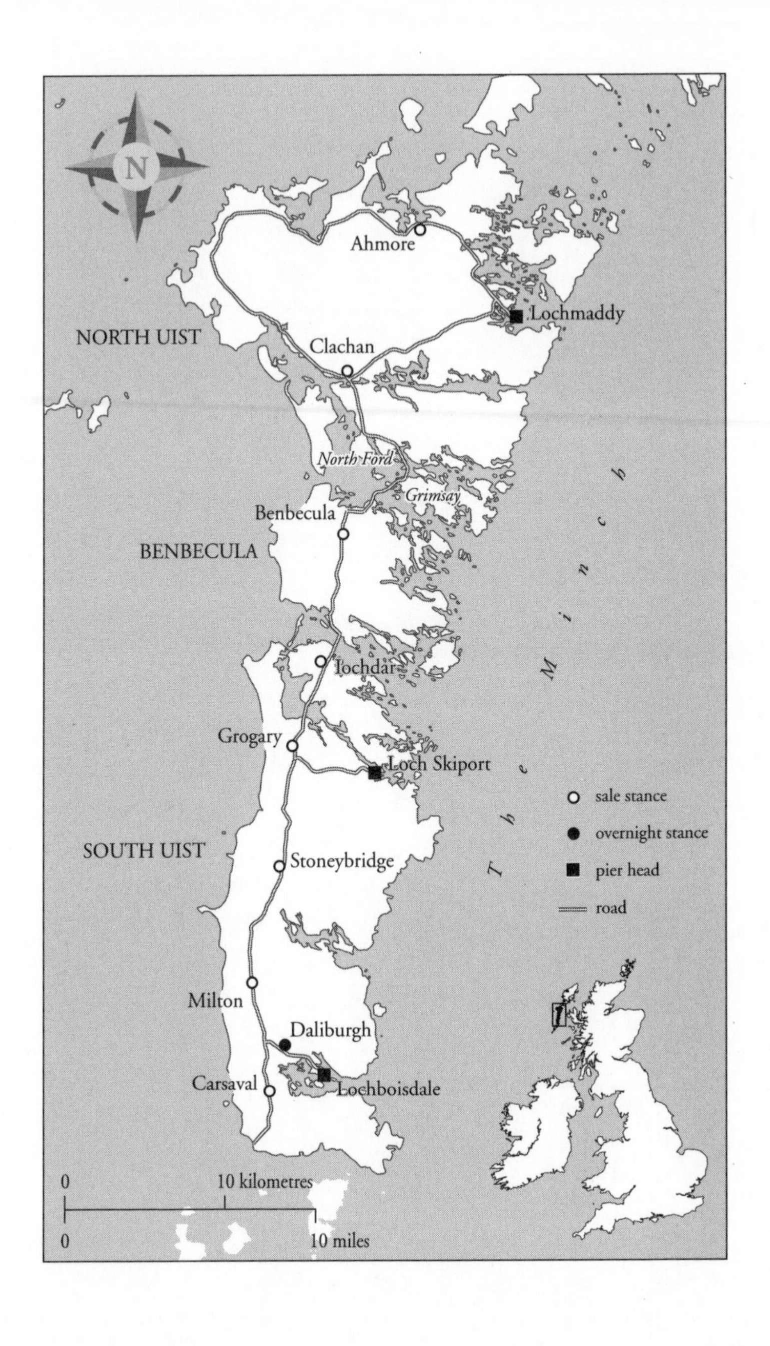
N
Ahmore
Lochmaddy
NORTH UIST
Clachan
North Ford
Grimsay
Benbecula
BENBECULA
Iochdar
Grogary
Loch Skiport
The Minch
sale stance
overnight stance
pier head
road
SOUTH UIST
Stoneybridge
Milton
Daliburgh
Carsaval
Lochboisdale
0
10 kilometres
0
10 miles

# 1

## *Beginnings*

The Highland bull gazed out across the Cromarty Firth from his vantage point above the road into Dingwall. Beside him stood a tall man, plaid over the shoulder, bunnet on head, stick in hand and dog at heel. I was looking for a campsite but the words 'Auction Mart' on a sign beside the road were irresistible. Being a farmer's daughter as well as an ever-inquisitive writer-in-a-campervan, I turned into the car park of Dingwall & Highland Marts Limited. The place seemed to be full of mud-spattered 4x4 trucks, which meant there was a sale. I pulled on wellington boots and walked over to the bull.

The bull and his companions ignored me. A plaque at their sculpted bronze feet identified the man as 'The Highland Drover . . . An Dròbhair Gàidhealach' and told me he had been here since December 2010. At the entrance to the mart building was a notice promoting 'The Drover Exhibition'. I anticipated a few glass cases displaying tattered documents and faded photographs, with maybe an 'artefact' or two for good measure. Perhaps it was worth a quick look round and maybe a plastic cup of coffee, if there was a cafe.

Double doors opened on to a light, spacious atrium. A babble of conversation rose from brightly covered tables where farmers, townsfolk and visitors were eating and drinking from sturdy white crockery and real mugs. The staccato, rapid-fire voice of the auctioneer in the adjacent sale ring collided with the bellowing of cattle from stock pens behind the main building and the sweet-acrid smell of livestock flavoured the temptations of good coffee and home cooking.

A flight of stairs led to a wide, circular balcony overlooking both cafe and sale ring. Round the walls I found, instead of the expected dingy memorabilia, a series of professionally designed, permanent display boards telling the drovers' story in words, illustrations, maps and photographs. From the first domesticators of cattle, through cattle reivers and drovers, to the auctioneers and truck drivers of modern times, the tale unfolded panel by panel and ended far across the world, with the drovers who took their experience overseas and developed the livestock industries of North and South America, Australia, New Zealand and South Africa. Any thought of a quick look round evaporated. I was hooked.

A colour video film, specially commissioned and professionally produced, showed footage of Highland droving country, complete with shots of a re-enacted cattle drove. It gave a flavour of the long journeys through the hills, an inkling of the hardships involved. And, if the presenter's commentary was occasionally lost in the din of a livestock mart in full flow, it only added to the authenticity. Further on, a small screen set into the wall showed, at the push of a button, slices of droving history from the Scottish Screen Archive.

I paused by a table full of labelled folders: Auctioneers; Marts; The Drovers; Sales in South Uist – a treasury of newspaper cuttings, photographs, letters and articles; a note to visitors explaining that this was just a sample from the

Highland Drover Project's main collection, which could be found in the Highland Archive Centre in Inverness, should I be interested. I would certainly visit the Archive Centre. More immediately, I had found a story and I wanted to know more. With no idea that this discovery would lead me south to Crieff and north to Lairg and, most significantly, across the Minch to the Atlantic edge of the Western Isles, I arranged to meet Kenneth (Kenny) McKenzie, Managing Director of the mart. Could he tell me about the Highland Drover and the exhibition and the collection and the people who carried out the project? He could.

In 2004, a public meeting was held in Dingwall and the Highland Livestock Heritage Society (HLHS) was founded 'to celebrate and honour the traditions and knowledge of the Highland drovers and their trade and to ensure that their legacy is preserved for the benefit of future generations'. The objectives were threefold – an archive of droving-related material, an exhibition and a commemorative sculpture. What's more, the venture was to be located not in a museum or a visitor centre but at the heart of the droving tradition in Dingwall's 21st-century, state-of-the-art livestock mart. It was a big challenge but the men and women involved were unstinting in their commitment.

According to the project's brochure, the Highland drovers were 'the heroes of their day, [who] helped build Scotland's livestock industry. And along the way, they supported and entertained countless communities throughout the region . . . [They] set out to gather the cattle and sheep they would go on to care for over several months, walking them through hundreds of miles of wild terrain to the great livestock trysts of central Scotland and beyond. Calling at isolated hill crofts, busy coastal villages and peaceful glens, they exchanged news and the payments on which many families in the Highlands and Islands depended for their very survival . . . Everywhere they went, the drovers were made welcome. These men combined skilful animal husbandry with physical toughness and the ability to negotiate fair prices and manage the complex and risky commercial structures of the livestock trade.'

The first task for the project team was to start collecting – first funds and then material for the archive and exhibition. David Henderson, an economist with a strong interest in droving history, brought useful experience of funding issues to his role as chairman. He also contributed a gentle sense of humour that helped balance the ideals of the project against the pragmatism needed for a fruitful outcome. Farmer's daughter Janey Clarke offered her services as archivist. Kenny McKenzie allocated her an office at the mart and the words 'Highland Drover Project' appeared on the door. Janey sat down at her desk and the work began. In response to a couple of newspaper adverts and a string of phone calls round the local farming community, letters and email messages began to arrive. The Highland grapevine spread the word. People came to the office with envelopes and packages bulging with memories of droving and the marts. This was their own history being recorded. Farms pass down from one generation to the next and, in many cases, there's a drover somewhere along the line.

Janey herself delved into books and magazines, newspapers and record books, typing out extracts and filing them. She put on her boots and walked some of the old drove routes. When I met her, it was clear that she had enjoyed every minute, despite the pressure of recording and organising such a large quantity of material. She had even spoken to a few people with tantalising memories of more recent droving days.

'My aunt introduced me to Kenny Stewart in Gairloch,' she said. 'His uncle, Kenny Macrae, was a drover in the first half of the 20th century . . . a larger-than-life character, a "big operator". He had a massive amount of land, partly owned and partly rented, but he was a self-made man. He bought cattle all up the west coast, from Ullapool towards Assynt, and went out to the Isles, driving and "buying, buying, buying" as Kenny Stewart said. That for me really brought the project to life.'

The two Kennys were included in the exhibition, as was John Cameron of Corriechoille in Lochaber, one of the greatest drovers of them all. He was another 'big operator', reaching the peak of his career around 1840. In those days, the droving trade was a mainstay of the Scottish economy and huge droves of migrating cattle filled the glens and mountain passes every autumn. Some of Corriechoille's droves were reputed to be over a thousand strong and seven miles long. More than two hundred years later the glens are empty and the passes silent but livestock still has an important place in Highland life and culture. Cattle and sheep now ride in shining lorries along fast roads to the marts of Dingwall, Aberdeen, Fort William and Oban. There's more denim than tweed to be seen at the sales today and more baseball caps than bunnets gather at the ringside. Descendants of the heroic drovers look round the exhibition, browse through the folders of old photographs and, like their forebears, they share news and stories whenever they meet at the sales.

One of those descendants found himself in the Drover Exhibition. On a display board labelled 'Uist to Dingwall by Ferry and Train', I found a photograph of a young man, cap on head, stick in hand, collie dog by his side. The year was 1954, the location North Uist. The dog's name was Tan and the young man was Ian Munro, a 20th-century drover employed by Reith and Anderson's auction mart in Dingwall (predecessor of the current mart). His task was to help walk cattle from the island sales to the boats that would ferry them across the Minch to the railhead at Kyle of Lochalsh. Now in his seventies and retired from his Ross-shire farm, Ian was still a regular at the mart, said Janey, and I was delighted when she introduced us. This kindly, quiet-spoken man was to prove generous with his recollections of what he described as 'some of the best days of [his] life'.

Not long after that first meeting, I settled in the Black Isle, not far from Dingwall. I became a frequent visitor to the mart and its cafe, which turned out to be one of the best in town. When time allowed, Ian and I would catch up on news over a mug of coffee. I heard many stories about his droving adventures in the Uists and Benbecula – the journeys out to the Western Isles with the auctioneer and the mart officials, the sales held in different parts of the islands, the bus that carried the dealers from one sale to the next, the local crofters who helped with the cattle and how it was a disadvantage that he didn't speak Gaelic. One day he spoke of his fellow drovers. There had been three brothers in South Uist. Ian remembered their names and the location of their croft but he had no idea if they were still living.

If I could get permission to make a smaller, travelling version of the Drover Exhibition, I could take it to the North Uist Show in August and maybe there would be people interested enough to share memories of the cattle sales of the 1950s and '60s when Ian was in the islands. I might even find his droving companions. I had no external funding but I did have my campervan and my writer's experience of gathering material. Lack of money had rarely prevented me from doing anything I thought worthwhile.

I consulted Kenny McKenzie, who suggested I put my proposal to the next project board meeting of the HLHS. In the official boardroom at the mart, at an imposing polished table surrounded by mostly unfamiliar faces, I perched on the edge of my seat. The board leaned back in their chairs and listened as I told them I wanted to trace the island drovers' story in a way that I hoped would add to their archive. By following a notional line from Uist to Dingwall, I would seek out individuals and record their memories or handed-down stories of the cattle sales, the sea crossings and the rail journeys. I would travel steadily, adapting the detail of my route as the journey unrolled. I would follow the trail from one conversation to the next, gathering stories like a drover adding cattle to his herd. At night I would sleep in my small campervan wherever the day's journey might end. It would be challenging but simple. I wasn't looking to the board for funding or even for their involvement (though any unofficial interest would be very welcome) but I did hope they would approve of my idea. The board asked questions, were encouraging – and gave me their blessing.

Janey was very pleased when she heard the news. She promised her full support. I must get in touch any time if she could help. Bundling up an armful of books about droving and cattle and marts, she insisted that she didn't want them back. They were mine to keep. I could add them to my droving bookshelf. And . . .

'You should have this,' she said.

She placed an irregular crescent of rusted iron in my palm, where it fitted almost exactly the curve of my lifeline. Its flat 'nose' broadened to a couple of centimetres before tapering away to a blunt point near my wrist. The upper surface was worn and uneven and pierced by three holes. I could see my skin through two of them, while the third was blocked with years of corrosion and dirt, but still a definite hollow. The underside of the object was smooth against my hand. Turning the crescent round, I recognised the outline of half a cloven hoof.

'It's a cattle shoe,' said Janey. Three nails, driven through the holes, would have held it in place. I was holding a piece of droving history.

'I found it in Knoydart, lying on the ground in front of me,' she said. How strange, and how fitting, but how on earth could she bear to pass on such a treasure? Perhaps it was like handing over the Olympic torch – my turn to keep the flame burning.

Back home, I packed the van: cup, plate, cutlery, kettle, pan; sleeping bag, extra clothes, waterproofs, boots; tape recorder, camera, notepad, pencils, laptop, mobile phone, one or two books; quick-cook porridge, long-life milk, dried fruit, lentils, tea bags, coffee . . . I had a bed, a cooker, a small water tank, a heater and a rechargeable supply of electricity.

The Highland drover of old would surely have been astonished. Essentials for him were his dog, his stick, a knife and the long, woven plaid that served as protection from the weather and provided safe keeping for his money within its folds. A bag of oatmeal, an onion and a horn of whisky would be his provisions. There was no shortage of fresh water in the Highlands, he would have the open hill for a bed, perhaps a heather-root fire if it got chilly. He carried a wealth of stories and songs in his head to while away the evenings and he'd sleep fine wrapped up in his plaid. A tough character, bred for tough living. His eventual destination would be the Crieff Tryst in the middle of Perthshire, so I decided to take my cattle shoe on a small pilgrimage before heading out to the islands.

## 2

# *A Small Pilgrimage*

I breathed a hole in the ice on the inside of the window. The tip of my nose was cold. Inside the sleeping bag, under the quilt, despite thick socks and a woollen hat, the rest of me was also cold. Outside the van, frost and moon glitter. A stag was bellowing from the hillside, challenging his own echo. I pulled on a sweater and lit the gas under the kettle.

The previous evening I'd abandoned the A9 just south of the Drumochter Pass and followed a single-track road into the hills, to this patch of flat ground beside a burn. Today, there would be the long swoop down to Trinafour, through to Tummel Bridge, up and over by White Bridge to Coshieville, with a wave to Schiehallion in passing; then by an airy whisker of road from Kenmore to Loch Freuchie, Amulree and the Sma' Glen . . . I'd be in Crieff by lunchtime.

The drovers used to come this way throughout the 17th and 18th centuries, bringing thousands of cattle on foot through Drumochter from the mountains and islands of the north to the great market tryst in Crieff. In 1723, one Bishop Forbes arrived at Dalwhinnie to find eight separate droves resting before the climb up through the pass. I tried to imagine those twelve thousand cattle filing past my campsite – 'the wealth of Scotland's mountains', as Dr Johnson called them when he toured the Highlands in 1773. He was right. Without cattle to sell, a crofter couldn't pay his rent; without cattle for her dowry, a girl had nothing to offer a prospective husband. Cattle imparted status. 'How many cattle do you have?' was a question of prestige.

Cattle were covetable and also very mobile. As early as the 14th century, a recognised droving trade of sorts was incurring customs dues on the export of beasts from Scotland into England. At the same time, the tract of land stretching from Solway to Tweed on either side of the border was a breeding ground of cattle thieves – the Border reivers – for whom the law was an irrelevance and guerrilla warfare a way of life. They wouldn't baulk at ambushing a drover and stealing his beasts, any more than they would hesitate to 'lift' a herd of good fat cattle from a neighbouring glen. They were tough, resourceful men with an intimate knowledge of their own ground; fine handlers of livestock as well as weapons; keen minds capable of calculating risks and taking action. Loyalty to their own family name came before allegiance to any national cause.

Droving through 15th-century Scotland was a perilous activity, hampered by many and various official prohibitions on cattle movement, as well as the dangers posed by far-from-official raiding parties. Despite the obstacles, a tenacious if sporadic cattle trade to England continued and the death of Elizabeth I in 1603 marked the beginning of a violent and irreversible end for the Border reivers. James VI of Scotland claimed the English crown as well as his own and, within seven years, he had effectively removed the border between his two kingdoms, along with the unruly Borderers. Recognition of droving as an essential part of the Scottish economy came in the 1670s, with orders that all drovers must carry identification. The reivers' descendants, inheritors of finely honed droving skills passed down through their fathers and grandfathers, could now swap infamy for a 'certificate of respectability'. Many surely did. With routes to the southern markets finally cleared of their ferocious ancestors, they, in turn, could profit from the wealth of Scotland's mountains.

For almost two centuries, droving flourished. After the defeat of the 1745 Jacobite rebellion, drovers were the only civilians permitted to carry arms. From early summer, they would go out to the islands and remote corners of the mainland, buying cattle and walking them gradually south in ever-increasing numbers. The Highlands of the 18th century would have been a land in motion. The straths and glens would be filled with a gathering stream of cattle: from Caithness and Sutherland, Moray and Aberdeenshire; from Mull and Islay, Kintyre and Argyll; from Skye and Lochaber; from Lewis and Harris, Barra and Benbecula and the Uists – all converging on a small Perthshire town, en route for the far-distant grazings of East Anglia and that greatest market of them all at Smithfield in London. It was a long, slow journey at ten or twelve miles a day and they, like me, would have frost and moon glitter on the way.

I found a handwritten letter in the HLHS archive, containing the handed-down story of a 15-year-old drover in the year 1860 or thereabouts. He was walking cattle from Glenlivet to Falkirk, which had, by then, replaced Crieff as the centre of livestock trading, being more accessible to English buyers. Sleeping out under the stars, the young man woke one morning to find his shoulder and arm frozen to the ground. The legacy of that cold night afflicted him for the rest of his life and, at the age of 95, his arm was still semi-paralysed. That young man would have been one of the last mainland drovers. Within a decade, railways and improved roads made his profession redundant – except in the islands.

Hot tea and breakfast had warmed me up. It was time to get on the road. The drovers would take much longer than my couple of hours to reach Crieff from here, especially if they were to stop and get the beasts shod at the Trinafour smithy.

Why put shoes on cattle? The early drovers would doubtless have scoffed at the notion. Their ways through the glens were no more than paths worn by generations of human and animal feet. But the 18th century brought General Wade to tame the intractable Highlands. His soldiers built roads along many of the main drove routes and the drovers complained that hooves used to softer ground were worn to bleeding by the hard, stony surfaces. It was a 600-mile walk from Uist to London and, the further south they went, the more hard roads they encountered. The cattle needed all the help they could get so they got shoes.

I could only speculate on what 'my' cattle shoe was doing in Knoydart, one of the most roadless peninsulas in the country. How did it get there? What hoof did it once protect, how far did it travel and how was it lost? Did the nails lose their grip in a patch of boggy ground or was there a wrenching-off against a sharp rock? Was the hoof itself damaged and did the beast become lame? And who made it?

Here in the tiny hamlet of Trinafour in the heart of Perthshire, with hundreds of cattle passing his door, an enterprising blacksmith saw a chance to make some money. It was said that one man could shoe seventy cattle in a day. The smith of Trinafour must have prospered, though with two shoes to each of four feet and every beast needing to be roped and thrown first, his profit was hard earned.

For the townsfolk of Crieff, the trysts meant chaos and uproar as cattle, drovers and dogs flooded into the small town from the surrounding glens and spread for miles across the open ground round about. Sales were agreed with the slap of a hand, the handing over of cash and, most likely, a dram in one of the many refreshment tents. There would be jostling for space, outbursts of fighting, shouting and barking and bellowing, stink, mud, and filth. And worse. The drovers – the more impudent of them – would be quite happy to take over a person's house and fireside, even the beds. Next morning there would be no potatoes left in the garden and the blankets could well have disappeared too. Formidable visitors indeed.

Hopefully, by arriving with my own accommodation, food and blankets, I would be a little more welcome than those unruly characters. The town was busy. A 21st-century 'Drovers' Tryst' was in full swing, devoid of either cattle or rampant highlanders. This event was primarily a walking festival to celebrate 'the life, work and play of the people who made Crieff the cattle-droving crossroads of Scotland in the 1700s', with guided walks into the Perthshire hills and glens, among the ghosts that surely linger. I wondered: if you stayed out alone under the Michaelmas moon after the guide and your companions had returned to the minibus and the warm, lamplit town, might you hear the steady tramp of cloven hooves?

The streets seemed to be full of people wearing colourful waterproof jackets, overtrousers, bright fleece hats, sturdy boots and neat backpacks. It was a far cry from the old days. Then, as the drovers crowded into town after weeks of hard, open-air living with their dogs and cattle in all kinds of weather, the animal reek of their thick, homespun plaids with overtones of peat smoke, sweat and, no doubt, whisky would swirl through the streets. By all accounts, the drover-dealers, who did the buying and selling and were often on horseback and hired others to walk with the cattle, were in the habit of wearing three coats: one elegant for formal occasions, another with pockets for all the cash they hoped to be carrying, the third an oilskin to keep out the weather.

I huddled into several warm layers and made my way to the church hall by the river, for a Drovers' Day of Gaelic songs, stories, poems and proverbs relating to drovers and cattle. Led by folklorist Margaret Bennett and a gathering of fine storytellers and singers, we listened and learned and joined in the songs. There was tea and good conversation. The one planned day unravelled into an impromptu weekend of shared enthusiasms. As everyone eventually went their various ways, I wrote the first of many phone numbers in my notebook. I had an invitation that would eventually take me on a diversion to Sutherland but, for now, it was time to return home and prepare for the journey west.

# 3

# *The Journey West*

Ian Munro, without whom this story might never have been written, grew up on an Easter Ross farm just three miles outside Dingwall. As a boy during the 1940s, he was more interested in cattle than school. When lessons ended for the day, he and his friend, George McCallum, would head for Reith and Anderson's auction mart in the centre of town, throw their schoolbags into the office and turn 'drover', walking the cattle and sheep from the sales down the main street to the railway station. In those days, most livestock from the mart travelled by rail, heading for Aberdeenshire and the east or maybe Perthshire, Stirling and the south and the local people were used to ducking into shops to avoid a herd of cattle or a flock of sheep as they went about their errands in town.

Reith and Anderson wasn't the only auctioneering company in Dingwall at the time. There was healthy rivalry with Hamilton's Auction Mart until the two amalgamated in 1992, forming what eventually became Dingwall & Highland Marts Ltd. Nor were Ian's employers alone in having an interest in the Uist sales. On the other side of the country, Thomas Corson's mart in Oban also held sales in the islands. For many years, the two companies had competed for the islanders' business but, by the 1950s, they were working together – sharing the financial burden, as well as auctioneers, clerks and drovers, and employing local helpers every spring and autumn for the week of the sales. An archive photograph might show an Oban clerk working alongside a Dingwall auctioneer, and a herd of cattle would be handled by drovers from both companies, never minding who paid them so long as someone did.

I asked Ian if he would point out the location of all the market stances in the Uists and Benbecula. I had the relevant Ordnance Survey maps – latest edition, scale of two centimetres to one kilometre, shiny pink covers. No, they wouldn't do. We needed to see the islands as they were in the 1950s. That was easier said than done. I discovered that all remaining stock of the 1959 OS issue (red cover, one inch to the mile, familiar companion of many a youthful hillwalking expedition) had been scrapped. I searched internet sites, antiquarian booksellers and even charity shops. I rummaged in boxes, sifted through piles, scrolled down lists. I found maps of Wales, of Middle England, of South West Scotland, of the Isle of Skye, all dated 1959, all with frayed and faded red covers. But maps of the Uists and Benbecula seemed to have vanished. Finally, I tracked down a set of originals in the National Library of Scotland and managed to buy photocopies of the three sheets we would need. These arrived rolled in a cardboard tube almost a metre in length and we spread them out over several cafe tables at the Dingwall mart, much to the entertainment of staff and customers.

There were eight sales, said Ian, held at strategic points throughout the three islands: one in Benbecula, five in South Uist, and two in North Uist – in that order. He traced the route of the droves, anxious to show me exactly where each sale had been held, pausing here and there to share recollections, descriptions and anecdotes. I scribbled grid references in my notebook while following his finger, hoping that I'd be able to read my own writing afterwards. I would transfer the information on to the more portable, if less authentic, modern maps that would travel with me. With those maps, a folder of old photographs and a scaled-down version of the Dingwall Mart drover exhibition, my intention was to begin at the beginning, in the last stronghold of on-the-hoof droving in Scotland, where Atlantic winds and waves made landfall after several thousand unimpeded miles of ocean and where some of the most sought-after cattle in the country were still reared. It was time to go. I had a booking on the afternoon ferry from Uig in Skye to Lochmaddy in North Uist and a display stand waiting for me at the North Uist Agricultural Show the following day.

I waved to the Highland Drover as I passed Dingwall Mart, remembering a recent telephone conversation with his creator. Perthshire sculptor, Lucy Poett, had faced high expectations when she received the commission from the Highland Livestock Heritage Society. The project team knew just what they wanted. The bull needed to be 'a good specimen of a beast' and the drover 'a strapping man, symbolising the virility of the trade'. They were particular about how the drover should be dressed, even to the details of his beard. A cattle farmer herself, Lucy recalled with evident pleasure their discussions about 'the shape of the bull's horns, the tail . . .' Her sample clay model was too lean, said the team. Could she add some flesh, perhaps? Of course she could. And so the Highland Drover steps out with a distinctly well-fed beast.

At the watershed above Achnasheen, I noticed an eastbound train, its two purple-and-silver passenger coaches creeping through the vast mountain-and-moor-scape. This was the line that once carried hundreds of island cattle in dozens of wooden livestock wagons hauled by steam-belching locomotives (and their diesel successors), from sea level at Kyle of Lochalsh to this high point of almost 200 metres and back to sea level at Dingwall on the edge of the Cromarty Firth.

I skirted the south side of Loch Carron where road and rail share a narrow strip of rockfall-infested shoreline. A few miles later, the sign for 'Strome Ferry (No ferry)' prompted a smile as usual. This crossing, at the narrowest part of the sea loch, used to be unavoidable for east–west travellers. Replacing the ferry with a road in 1970 probably seemed a good idea at the time but it has caused problems ever since for users and operators alike. Frequent landslips cause closures, some lasting weeks that have occasionally turned into months.

Today, the road was clear and, half an hour later, I was crossing the bridge to Skye, with a glance at the Kyle of Lochalsh pier far below. Throughout the 1950s, this was where the drovers boarded the regular mail boat, *Loch Mor*, which took them up the Sound of Raasay, round the north end of Skye and across the Minch to the Outer Isles. Then, on 15 April 1964, everything changed. David MacBrayne opened a new route with a brand new ferry and brought to an end the *Loch Mor*'s career in the Minch. From that day, the *Hebrides* took on what became known as the 'Uig Triangle', working between Uig, Tarbert in Harris and Lochmaddy.

The new boat had hydraulic ramps that could be raised and lowered according to the state of the tide. This was a huge innovation, which meant that almost the entire Western Isles archipelago had access to a car ferry. South Uist and Benbecula had been joined by a bridge across the South Ford during the Second World War and the North Ford causeway between Benbecula and North Uist had just opened in 1960. What's more, the *Heb'* could take cattle trucks across the Minch. Soon the island cattle would be travelling in style like their mainland counterparts and there would be no need for drovers to walk the beasts from the sales to the piers. Would I be in time to find people who remembered? Maybe, so long as I caught the boat!

Sligachan and the Cuillin flew past with the briefest of greetings. Portree came and went with hardly a glance and the road to Uig was mercifully uncluttered. One final pause above the village curled round its circular bay, then I was swooping down the last steep hill to the ferry terminal with minutes to spare. There were tickets to collect, a queue to join and activity among yellow-jacketed CalMac officials that signalled the imminent departure of the ferry – *Hebrides* the third. From the rail of the open deck, I watched Skye grow smaller. In less than two hours, I would be in North Uist; the drive across Scotland had taken just under three. My journey bore little resemblance to that of the Dingwall team almost 50 years earlier in 1965.

Ian Munro was an old hand at the job. He had been crossing the Minch twice a year since the 1950s, employed by Reith and Anderson as a drover. In his early droving days the sea voyage began at Kyle of Lochalsh, and the Isle of Skye was no more than a passing coastline.

'The *Loch Mor* left Kyle, went way across to Scalpay, then it went from there to Tarbert Harris, then it moved down to Lochmaddy, then down to Lochboisdale in South Uist. That was about twelve hours but it was more than twelve hours by the time we left Dingwall . . . I would be eighteen or nineteen when I first went out there and that lasted about ten years. The job I was doing wasn't really required once they took haulages out, the transport . . . There was only two of us [drovers] went out from Dingwall but there'd be another six employed [on the islands] . . . All the locals that were working, Gaelic was their first language. We did not have Gaelic, much to our disadvantage!'

In 1965, young Kenneth McKenzie was a trainee auctioneer on his first working trip to the Uists. In 2013, the year of my journey, he would be on the verge of retirement as Managing Director of Dingwall & Highland Marts Ltd (Reith and Anderson's successor) after a lifetime of selling livestock. 'We left Dingwall on the Friday morning and we spent the whole day travelling out to Uist. It was two hours to Strome Ferry. You had to allow an hour for waiting at Strome Ferry, then single track road round to Plockton, another hour at Kyle, then of course the roads through Skye weren't that great.'

Another newcomer in the team was George Tait, who had recently joined the Dingwall firm as Cashier and Company Secretary. His previous experience had been in East Lothian. This would be a whole new adventure for him. 'What I was coming to, I had no idea! Coming from East Lothian and going out there, all these hairy animals – mostly cross-Highlanders – was quite a shock to the system! But it was great experience and I'm glad I was there.'

The ferry churned steadily westwards.

Loch nam Madadh (Loch of the Wolves, named after three sharp-toothed skerries at its mouth) was surely too narrow, too shallow, too rock strewn for a ship this size. But Hebridean skippers know their job. Inch by inch, the boat approached the pier. Ropes were thrown and tied, the ramp went down and the village of Lochmaddy looked on as *Hebrides* discharged her cargo of vehicles into the sunlight. We had arrived.

At the head of the pier, modern cattle pens and a small mart building were instant confirmation of why I was here. It was tempting to stop and look round but first I needed somewhere to spend the night. The showground was on the opposite side of the island from Lochmaddy. One 'main' road circles North Uist, with little to choose between the northerly or southerly route, save that the first consists entirely of single-track roads with passing places and arrow-straight sections liberally sprinkled with blind summits and hidden dips, to which local driving habits add an extra thrill. Half of the southerly option allows two vehicles to pass without stopping. I turned south.

My road was a grey thread through a watery landscape where intermittent contours meandered among peat bogs, clustering here and there into low hillocks, many of them topped by ancient heaps of stone. Between the hillocks and the water, dark tiles of cut peat were laid out to dry. At intervals, rough tracks led off the main road into the bog, punctuated by small heaps of already-dried peat. Here and there, a crooked finger of sea probed so far inland that it seemed only a few yards of dry(ish) ground prevented a commingling of Minch and Atlantic. At the southern edge of the island, I turned west into a different country. This was the machair land, the cattle land, fertile and well drained, with sweet summer grazing and dunes that gave shelter from Atlantic winds and storms.

I had reached the true beginning of my journey. Since Ian's

droving days, market stances and piers had fallen into disuse, roads had been realigned, new houses built, people had aged. In the morning, I would set up my display in the showfield among the cattle and sheep, the home baking and garden produce, hoping for interest, hoping for stories. I fell asleep to the rasp of corncrake and the thump of ocean on sand.

# 4

# *Searching for Drovers*

'Simon, Neil and Charlie Campbell,' Ian had said during our map study in Dingwall. 'The family croft was near Lochboisdale.' The brothers had shared the days of walking island cattle to the boats and I wanted to find them.

My day at the North Uist Show led to many enjoyable recording sessions at many kitchen tables. Local people expressed enthusiasm for my project, invited me into their homes, plied me with food and cups of tea and were usually still sharing stories when I left, often several hours later. I quickly learned not to make more than one appointment at a time for there was no telling how long a meeting might last or which direction I might be sent in next. An hour (my usual estimate of how long I ought to impose on anyone) was rarely enough and the flexibility of campervan living meant I could go with the flow. It may have been inefficient time management but the harvest was plentiful as a result and I found the gleaning went on for many months after I had returned home.

The first of these welcoming households was that of Catherine and David Muir, who live in Benbecula, hardly a stone's throw from the starting point of the drovers' week of sales. They both have busy professional careers but crofting in general and on their own account has always been at the heart of their lives. For me, their knowledge of local history and the crofting community was like finding gold. For many years David was the local representative of the Scottish Crofters Union (now the Scottish Crofting Federation) and Catherine's involvement with the Benbecula Historical Society was the key to a hoard of photographs relevant to my search. I left with several notebook pages of names, addresses and phone numbers and instructions, above all, to contact John Macmillan the very next day. On a Sunday? Yes of course, just do it, urged Catherine. So I did.

I found the Macmillan family croft near the western end of a narrow side road that eventually wandered into the machair and dwindled into a track among the dunes of South Uist. John was expecting me and we were soon surrounded by photographs and papers as he generously shared his wealth of local knowledge. He told me his great-great-great-grandfather used to keep 600 cattle on the rugged east side of the island. I brought out an archive photograph of a sale in progress and he picked out his grandfather in the crowd. Had I noticed a big stone enclosure as I left the main road? That was Milton Fank where the sale used to be held, though it had been much altered and added to over the years and was still in use for handling the crofters' livestock. The voice recorder caught story after story. Somehow a mug of tea and a plate of biscuits appeared at my elbow. Time passed.

Back in the van I spent the evening typing up my notes. Suddenly two names caught my attention: Neil and Simon Campbell. I had found Ian's drovers.

* * *

Neil Campbell's wife answered my phone call. Her husband was digging in the garden, she said. Then: '*A bheil Gàidhlig agaibh?*'

'*Beagan* (a little),' I replied, which was apparently enough for our brief conversation to continue in her native tongue. I felt unjustifiably complimented; my Gaelic was certainly not adequate for the discussion I had in mind. I heard her call to Neil that there was a woman on the phone wanting to talk with him. Yes, I could come to see him any time at all for wasn't he retired now? Just come, any time.

By the time I arrived, the gardener had abandoned his digging and was waiting at the door. Looking two decades younger than his 82 years, Neil welcomed me with a firm handshake. I recognised him from a life-size photograph I'd spotted during a quick visit to South Uist's Kildonan Museum. A tall, good-looking man, interrupted at his peat cutting, looking directly at the camera. Neil was not impressed to hear of his unsought fame. There was a pride in him nonetheless when he settled into a chair beside the kitchen stove and began telling me about the part he had played in the Western Isles' livestock trade. Any romantic notions of droving cattle in idyllic surroundings were quickly dismissed as he recalled the exhaustion of long days in difficult conditions and frequently poor weather, with insufficient helpers and uncooperative animals. The drovers' skill had been vital to the sale and transfer of island cattle to the mainland and he assured me that, whatever information I might get from other people, only those who had actually walked with the cattle knew the full story.

I heard about the cold and the beasts running off and no fences and the dogs and the pay of £8 or so for the week's work and the long days out in the weather . . .

'Some of the cattle were easy enough,' said Neil, 'but you would get the awkward ones, oh yes, trying to get away all the time, just trying to get away. They didn't have time for grazing . . . There's something else as well. You were without food all day. Well, later on in years there were vans going round but that time, when I started out, no nothing . . . I must be the oldest at this end of the island that was involved with it. There used to be another five of us.' His brother Charlie was no longer with us, he said, but Simon still lived in the family home near Lochboisdale. Was I going to see him? Yes – and I'd been given instructions for finding the house.

'What I told you, I was there,' Neil emphasised as I finally switched off the recorder and got up to leave. Indeed he was and I felt as though he had taken me there too, into the weather and the sounds and the smell of it all.

* * *

Simon Campbell was waiting for me as my van bumped gently along a grassy track towards a plain grey house with smoke streaming from the chimney. A welcoming handshake and an ushering into the warmth, where tea and biscuits were ready on a tray in the kitchen. Once again, I was made me to feel a special guest.

There was a clear family likeness between the two brothers but each cast a different light on their droving days. Neil's tales of hardship and misadventure turned to comedy in Simon's telling of them. Hardly had I set up the recorder than we were out walking the roads with the cattle. Gesticulating, eyes flashing, he brought each scene to vivid life. At several points, he jumped up from his chair, miming the effect of island hospitality on a mainland cattle dealer or a beast leaping over a fence with men and dogs in pursuit. His voice rose and fell to suit the drama of the occasion, the rhythms of his native Gaelic underpinning his answers to my English questions.

The result was almost musical. Gaelic was the first language of nearly everyone who helped me. An occasional slight pause between my question and the response hinted at the moment it took to shift their instinctive answer into English for my benefit. But Simon laughed when I told him how Ian's lack of Gaelic had been a 'great disadvantage' and apologised for my own inadequate grasp of his native tongue. That reminded him of another story . . . I brought out my file of photographs and we were soon putting names to faces among the dealers.

'That's Andrew Hendry from Stirling there,' said Simon. 'And then there was a Mr Binnie – the one on the right, I think. Then the one on the left is Ian Oswald . . . Some of them were staying here – was Neil telling you? . . . They were sleeping down in the room. Bob Love and Tom Adams and Willie Hendry and Andrew Hendry. Two double beds down there, just two in each bed. They were staying here until my mother got old and she got good lodgings for them and they were very happy.'

A tingle ran up my spine. I was sitting by the fire next to the room where the cattle buyers had slept and a tea tray was waiting for me in the kitchen where they had eaten their meals. It all suddenly felt very real. I mentioned that the Dingwall team had stayed at the Lochboisdale Hotel. Oh yes, Simon remembered, and the Oban folk stayed 'at our neighbours down here. They would phone to say they would be home at eight or nine o'clock. Maybe the sale was finished at six o'clock but there was a couple of jars going after that and sorting the cattle . . .' His mother, wanting to have the dinner ready as soon as her lodgers got in, would put the potatoes on to boil at eight o'clock, but 'they would be nine o'clock and after nine,' said Simon. 'She was worrying about the potatoes getting cold.' Her neighbour, wiser to the ways of these cattlemen, waited until they came through the door before putting the potato pan on the stove. 'And then there was drams going in the house down there,' said Simon. 'They were getting good feeds and they were happy.'

I thought there must have been a lot of fun as well as a great deal of hard work. 'Oh yes,' said Simon, 'down at the pier and watching them going on the boat and this hullabaloo going on. There was never a dull moment. You would get a wee dram now and again. That was keeping morale going. There was no such a thing as tiredness, no.'

* * *

The van, perched on the very southern edge of South Uist, swayed in a stiff breeze. I sat in the open doorway with a cup of coffee, watching a fly-past of gannets and listening to the day's recordings. The Campbell brothers had both been eager to share their memories and, as their voices drifted in the bright air, I realised that each of the three drovers had spun me a different colour of the same yarn. Ian, the son of a mainland farmer, found wonder in sights and sounds that Neil and Simon had known all their days. For them, the adventure began with the arrival of the twice-yearly cavalcade of strangers in their midst and the welcome break in routine that the sales provided. Their talk was of the characters, the disasters and funny incidents along the way that enlivened the journey. They were on familiar ground, they knew the cattle and the people and of course they were at ease in their own language.

'At every second croft you would hear the weavers at work,' Ian had said and I could almost hear the soft thud of cattle hooves, the clack-clacking of the looms as crofts came and went, the constant booming presence of the Atlantic. By the time I reached the islands, only the last of these was unchanged. Ian's was the thread that had led me here and now I held two more. My task was that of the weaver.

# 5

## *Circles on the Map*

BENBECULA, IOCHDAR, GROGARY, STONEYBRIDGE, CARSAVAL, MILTON, CLACHAN, AHMORE

Eight sales over six days. The list was like a roll call, repeated without hesitation and always in the same sequence wherever I went. Like the ocean and the wind, cattle were a fundamental part of life in the islands and the droving teams were as much a sign of the changing seasons as the arrival and departure of migrating birds. I had marked the location of each sale with a circle on my map and set myself the task of locating and photographing them all.

### SATURDAY: BENBECULA, IOCHDAR, GROGARY

On Saturday morning, drovers, buyers and auctioneers travelled the 26 miles by bus from their lodgings in Lochboisdale to the northern end of Benbecula. Stansa na Fèille is still marked on the map but today a large sign beside the road redefines it. 'Market Stance Waste Transfer Station' is busy with the comings and goings of council bin lorries and loud with the rumble of machinery and engines. It's hard to imagine the queues of men and cattle waiting to get into the sale and the bellowing of young beasts that sent young Catherine Muir to hide under the bed with her hands over her ears. She couldn't bear to hear the distress of calves separated from their mothers for the first time in their two or three years of life and, to this day, she won't let David put cattle on their croft.

'The first sale would be finished by about half past eleven or midday,' said Ian. 'Now, all the sold cattle were grouped together and the likes of myself and a few more, we walked that cattle along the road.'

'There would be about three or four locals,' said Simon, 'and two or three from Reith and Anderson's of Dingwall, two or three from Corson's of Oban.'

'Oh but driving these cattle along the road there,' said Neil, 'no fences, a couple of townships on the east side of the road, five or six townships on the west side of the road. Every beast was wanting home.'

'You had to run in front of them,' said Simon, 'trying to keep them back. And the other ones were wanting the other way, down the north and south, and swimming the lochs.'

'Most of those animals were led to the sale on a halter and they were pets,' said Ian. 'And when you put them all together, if they broke away, it's home they wanted to go and it took a good dog to turn them.'

'If you put a dog after them, they wouldn't budge,' said Neil. 'They'd stand there, looking at you or looking at the dog. A good strong dog would give them a bite on the nose, then they would shift.'

A bite on the nose would be quite a shock to a young beast that had spent its first two years as one of the family. With just two or three cows on a croft, each new calf would be petted and fussed over by the crofter's children. Calves were taken from their mother not long after they were born, as the cow's milk was needed to make butter and cheese for the family table. The calves got the skimmed milk that was left over and they had to learn to drink from a bucket. This didn't take long. The sucking instinct was so strong that they would latch on to a couple of human fingers and follow them down into the milk. Gradually the fingers were drawn away and the calf would be drinking by itself. Before long, the clanking of a bucket would set them bawling, eager for their breakfast, and their heads would plunge straight in with no need for guiding fingers. No wonder Catherine Muir was upset by the sound of young beasts crying for their mothers.

While the drovers and their dogs were persuading homesick cattle to stay with the main herd, the rest of the sales team and the buyers would be heading back down the road on the bus, to begin the next sale at Iochdar in South Uist. Many hours and many miles on foot lay ahead of Ian and his companions before they enjoyed the luxury of wheeled transport again. The drove would gradually settle to a steady pace until they reached the southern edge of Benbecula, where the next obstacle lay in wait. Having made their sales and pocketed the money, the island's crofters were ready for a bit of socialising at the Creagorry Hotel.

'I remember,' said Neil, 'when we used to go past, everybody in there came out. They were half drunk. It was: "That's the beast I sold!" and "That's the beast I sold!" and "That's the better beast!" And we were shouting them to go back. No way, no. They were just standing there, right in front of where the cattle were supposed to go. And the cattle would scatter all over.'

There would be more chaos and shouting, barking and

running before order was restored. Peace was short-lived as it was less than a mile to the South Ford Bridge. The bridge was both blessing and curse for the drovers. Built in 1942 for wartime traffic, it avoided the dangerous foot crossing of the tidal gap between the Benbecula and South Uist but was viewed with great suspicion by the cattle.

'Driving the cattle one time there,' said Neil, 'it was just murder. Two of us – we arrived at the bridge and it was slashing down rain and strong wind. Not a beast would go across.'

'The ones that was coming from Benbecula,' said Simon, 'they've never seen the bridge (it's a causeway today, as you know). Oh they weren't wanting to go on the bridge. They started scattering again, with them too scared to go on the bridge. Oh that's how it was.'

'By the time we arrived at Iochdar in South Uist,' said Ian, 'the second sale was almost completed and our drove of cattle was increased. And then we went on to another sale which would be starting maybe about four o'clock. That was at Grogary and that was the last sale of the day. By then, we would have maybe two hundred and fifty cattle in total and they were put into a big fank, like a stone enclosure. And they were left there until Sunday morning.'

Up to this point, all the cattle from the sales were grouped together, regardless of who had bought them but each beast had been marked with a daub of paint according to its mainland destination: green for Kyle of Lochalsh, orange for Oban. It would also carry the buyer's own mark. The marking was a skilled job, as I learned from George Tait who had acted as cashier for the Dingwall team in the Uists.

'You can get a huge number of marks on an animal,' he said.

'I know them all: left shoulder, top shoulder, left rib, top rib, left boss, right boss and so on. The buyer had the same mark year after year. You just knew them without asking because it was the same buyers that came year after year after year. They got a wee scissor clip as well – if the paint rubbed off, there was always a scissor clip. Of course these animals had plenty hair. It was no bother to put a scissor clip on them. But the men who did the scissor clip – they were good at it, you know, it wasn't just everybody could do a scissor clip.'

'So, there was a paint person?'

'Oh yes and a scissor clip person as well,' said George.

### Sunday: loading cattle at Loch Skiport

There was no long lie for anyone on Sunday morning. After another early start and the long bus journey from Lochboisdale, 'Out we went again,' said Ian, 'and we drafted the cattle into two lots: the Oban cattle and the cattle that went to Dingwall. They were known as the Kyle cattle because Kyle of Lochalsh was where they were landed.'

The cattle boats would come to the pier at Loch Skiport on the east side of South Uist to collect beasts from the Saturday sales, which would be walked the five or six miles from the overnight fank and loaded for the sea voyage to the mainland. At that time, Ian pointed out, 'ninety per cent of North Uist was Protestant and ninety per cent of South Uist was Catholic.' Loading cattle on to boats on a Sunday would not have been allowed in North Uist, whereas there was no problem in the southern island.

'There were two boats waiting in Loch Skiport,' said Ian, 'and whoever had the greatest number of beasts – I think it was usually Oban – their drove went away along the road first, followed by the second one. Then they were loaded on to the boats. All the Saturday's cattle was cleared off on the Sunday.'

'See the cows from the machair in the west,' said Neil. 'When they got to a soft place, they were just walking right through it, because they'd never seen a bog before.'

'Big cattle that came from parts of Benbecula,' said Simon, 'when they would go off the road, they would half bog! Trying to keep them going in case they would get stuck.'

I went to Loch Skiport. The road can hardly have changed since the 1950s apart from a skin of tarmacadam, winding through lochans and rocks and bog. Many times I thought I had reached the sea, only to find it was yet one more freshwater illusion. Here of all places, it was possible to imagine how it must have been: the shuffle of cloven hooves, the occasional splash and bellow as a beast miscalculated the terrain, followed by shouting and barking until the poor creature was back on safe ground. And while all this was going on, the rest of the drove had to be kept on the move, ever eastwards to the edge of the island, where the boats and the Minch were waiting.

At last I reached the end of the metalled road. A wide stone-and-earth track disappeared through a cutting in the rocks and plunged down a steep bank through two hairpin bends to the sea below. Having endured unfamiliar roads, perilous bridges and treacherous bogs, a hundred or so nervous cattle would find themselves crowded together here and driven through the gap. A stout drystone wall along the seaward edge of the track would prevent a panic-stricken stampede into the water. At the bottom a wooden pier. Here the boat would be waiting.

'They were hardy boats yon time,' said Simon. 'They weren't like the floating hotels you've got today, these ferries. Carrying cargo and everything, yes.'

A photograph taken at Loch Skiport around 1898 shows a busy landing place with a ship tied alongside, a cargo of stores in barrels, a gathering of people who look like holidaymakers and several horse-drawn carts waiting at the head of the pier. In 1901, someone made a postcard from the photograph and added a caption describing how the place was sad now that the passenger ships no longer came from Glasgow; the walls were in need of repair, and the remains of the pier as they emerged from the sea were like the fingers of a drowning man.

By the time I arrived the pier was a skeleton but it was more than I had dared hope to find. The tide was low and the drop was fearsome.

'A lot of your loadings depended on the tide,' said Ian. 'A medium tide made the loadings far easier.'

So, on top of everything else, there would be the tides to consider. Tides and the boats arriving and keeping the cattle steady along the way and not getting stuck in the bogs. Then getting the first drove loaded and the boat away in time for the next boat to arrive and the next lot of cattle to be driven on board, before the water level dropped too far.

'Loading cattle there,' said Neil. 'Nothing to hold them back, no pens, no nothing. Och, it was just horrible. If they broke away from the gangway, you had to watch. They would go straight for you. Hard work, hard work, aye. It was dangerous.'

There was danger in the sailing too. Once clear of the sea loch and beyond the shelter of the island, there was the Minch and the weather. The cattle had a long way to go yet. For the drovers and their dogs there was the bus ride back to Lochboisdale and a brief Sunday evening of rest. Oh, but not always the bus, Neil told me, laughing.

'Once it was a big van from Benbecula they had for the drovers. Oh, what a size of a van. No windows, no nothing. Well, we went in with the dogs and the drovers from Oban and Dingwall went in and the dogs started fighting. It was dangerous, dangerous. Anyway, they got it stopped. We were getting on the bus at times but that time it was the Big Black Maria.'

### Monday: Stoneybridge, Carsaval

'On the Monday morning, we all went out to Stoneybridge,' said Ian. 'When the sale was finished, the cattle were all put together and they were walked to Daliburgh and they were left there.'

The Stoneybridge sale in the archive photographs was a maze of walled enclosures busy with buyers and sellers and the bus and the cashier's car were parked on the road alongside. I went in search of the stance that Ian had circled on my map. All that remained was a triangle of rough ground between the old road and the modern A865. I parked and walked into the middle of the triangle. I guessed that most of the stone walls were buried under the new road and that the rest lay beneath the rough turf where I stood. Demolished or decayed, the result was the same and I felt keenly the extent to which past lives can disappear without trace. It wasn't the first or last time I would have such thoughts and they always reminded me of the reason for my journey. Without the guidance of those who had been part of the story, I would never have found Stoneybridge market stance. Places and people lose their purpose and eventually disappear and are forgotten – unless somehow they are recorded.

It was about eight miles to Daliburgh, where the overnight holding area was a narrow strip of ground bounded by a loch on one side and the road to Lochboisdale on the other. This was Corson's Park, named after the Oban auction company which had erected a post and wire fence along the roadside to help contain the cattle.

'Corson's Park,' said Neil, 'yes, what a place. We'd a job getting them out of there. We were getting bogged in. Och, that was no place for cattle at all. And one night there was a lot of cattle in there and they had a local man to look after them. Next morning there was no sign of any of the cattle. They were away, oh miles. We had to start gathering them . . . He must have gone away and slept.'

'While the drovers were walking with the cattle,' said Ian, 'the buyers and auctioneers and office staff, they went away to a place called Carsaval. The afternoon sale was held there and then that cattle were brought up to Daliburgh.'

At Carsaval my map circle seemed to be located on a bare hillside and I held out little hope of identifying it on the ground. Yet, when I arrived at the spot, there was a neat group of well-maintained stone-walled enclosures. Clearly the local crofters had adopted the old sales stance for their own use and they had certainly created a better place for handling livestock than the scene Neil remembered with a touch of grim humour.

'Och well, no fences no nothing,' he said. 'The cattle, they were just going down to the loch and swimming across – twenty or thirty of them in one bunch. There was a dealer there – he was from Skye, I think, but I can't remember his name. Anyway, he had a wee terrier that was great with the cattle. I remember one beast went from Carsaval to Daliburgh there and it went down to the loch. The wee terrier went after him and got on his back. It turned him back, it did, yes.'

### Tuesday: Milton and loading at Lochboisdale

'Tuesday morning there was a sale at Milton,' said Ian. I had a photograph showing that sale in progress, a house and byre in the background, flat croft land stretching far away to the west. John Macmillan had told me that the house belonged to the Macleod family. John Roderick Macleod, known to everyone as JR, still lived there and he had been involved in the droving. I should go to see him. Another phone number went into my notebook and, a couple of days later, I found JR at home in his bachelor croft-house.

We sat at the kitchen table with its view to the west and the old family home visible through the window. Yes, he had been out with the drovers, employed by Corson's of Oban like the Campbell brothers, though he was several years younger than them. Born in 1939, he said, left school in 1954, never married and worked on the family croft. He and Simon Campbell had always worked together and still did, helping each other at the hay and the peats. He had just fifteen head of cattle on the croft nowadays. No car. The tractor was his means of transport. What he remembered best were the sales themselves. The photograph? Oh yes, he remembered that scene and 'See the auctioneer – he was standing on a special block to give him extra height above the crowd.' JR said we would go over to the fank and he would show me where the sale ring used to be and maybe we'd find the block too.

Amongst the confusion of pens and passageways, JR pointed out the original walls and I touched them as if to come closer to the people and the animals who had been here all those years ago. There were still cobblestones underfoot in the old sale ring. JR was rummaging in the weeds. Then, success! He had found the auctioneer's stand – a concrete block about 12 inches high. Now I was standing in the middle of the archive photograph, imagining the stick-wielding buyers and eager crofters and a frightened young cattle beast that slipped and slid on the stones as the auctioneer urged up the bids.

When we parted company, JR gave me one more vital thread, though I had no idea of its significance at the time. He said I should go to see his brother, Ewen Macleod, who had gone to the mainland as a young man to work for Corson's Auctioneers in Oban. He still lived there. I'd find him if I went to the mart.

* * *

A few weeks later I joined Ewen Macleod at the annual show and sale of the Highland Cattle Society at the Oban livestock mart. Yes, he would be glad to tell me about his work as a drover in the islands but first we needed to watch a few beasts going through the ring. Prices were terrible, he said.

Ewen had left home in 1957 at the age of 19, telling his mother he was going to Oban for the weekend. Instead of returning to Uist, he got a job on a farm and five years later he was a full-time employee with Corson's. He was one of the their team at the Uist sales and no doubt his mother was glad to have him home for a week.

'My job was to go out and do all the painting on the cattle,' he said.

I had found the paint person.

'There could be thirty marks,' he said. 'Back of the head, shoulder, right shoulder, left shoulder, across the back, right loin, left loin, kidney, right hook, left hook, tail head, right plate, left plate, and so on. When you got that into your head, you didn't bother looking at the book. It stuck in your head. Every buyer had his own mark, year in year out. Andrew Hendry would be right loin and Andrew Binnie I think was left loin. I would see them bidding and I would get the stick ready, thinking, "He's going to buy this one anyway." I would have the stick ready with the yellow on it, just to put on. Nine times out of ten you were right. But then yon time, he would drop out at the last minute and then you had to change your stick or the place, you know. It's amazing, the memory. Then you took the cattle into Oban on the cargo boat. I think the last cargo boat was the *Loch Broom*. And all the cattle came in on that boat into Oban and then you put Andrew Hendry's to one side. Different colours, different pens.'

Did he do the scissor clip as well?

'Yes and there was a knack of doing it, you know. It's got to be deep. There's a way of holding the scissors – an edge. Well, if it's flat, it's no use – it won't leave a mark. Och, I enjoyed it and folk at home would feel so proud of you, seeing you mark the cattle. You belonged to the place.'

* * *

The drovers would gather the cattle from the Milton sale and walk them to Daliburgh to join the animals from Carsaval. Now that the sales were finished in the south end of the islands, the whole team – auctioneers, buyers, clerks and all – were expected to lend a hand. 'Oh yes,' said George Tait, 'I might have been a clerk but, once the book was closed, you were one of the boys. You know, get the leggings on and get them chased. There was good camaraderie amongst the buyers and that. Of course, we were all called the drovers. You could be a buyer, an auctioneer, a cashier – we were just known as the drovers.' And they would often find themselves with one or two further assistants. Even as a small boys, Ewen and JR Macleod would be keen to join in the excitement of the sales.

'I didn't like school anyway,' said Ewen. 'I'd say how many days to the sale, counting on your fingers, getting closer and closer. A wifie there, I think a huge tent she had, with teas and that. Oh it was lovely, something different, you know – then I must have been ten or something.

'You used to follow the drove on the road and I went so far going up to Daliburgh. The drover says you'd better go back now in case you get lost. Well, all the buyers were behind the cattle and they were – mind the two-and-six, half a crown? – they were rolling them along the road to me and me gathering them up. Oh, I wandered home, counting how much I had. And many's the time we took cows on a halter from the crofts out there to Lochboisdale, to go on the boat for Oban. We thought it was great. If you got a pound, oh gee whizz, I've got a lot of money, you know.'

The cattle waiting in Corson's Park were sorted for Oban or Kyle according to their paint marks, then they were driven the remaining few miles into Lochboisdale. Loading cattle here was no easier than at Loch Skiport, said Neil. 'At low tide, there was a slipway under the pier and loading cattle there – you couldn't walk because of the seaweed. Well, the cattle were sliding there as well. If a beast broke away – aye, it was dangerous, it was really dangerous.'

One time at high tide, Kenny McKenzie remembered, 'They crowded together on the Lochboisdale pier and we were all holding them and they burst out at the side and half a dozen of them went over the edge of the pier into the water.'

By late Tuesday afternoon, the laden cattle boats would be on their way to Oban and Kyle. Drovers, dealers, auctioneers and clerks said goodbye to the hospitality of Lochboisdale, then it was 'back on the bus again, to get the ferry across to North Uist,' said Ian.

Up through South Uist, past Daliburgh of the bogs, past the sales stances at Milton, Stoneybridge, Grogary and Iochdar, over the wartime South Ford Bridge to Benbecula and past the Creagorry Hotel. At Stansa na Fèille, site of the Benbecula sale, the coastline of North Uist would come into sight. Down the last hill to the end of the road, a slipway and the North Ford: four miles of water or sand depending on the state of the tide, separating the two islands. Twice in every 24 hours, the sea withdrew (as it still does) eastwards into the Minch and west into the Atlantic, like a biblical parting of the waves. Crossable by boat, on horseback or on foot, the place was as fickle as the west-coast weather, no matter what manner of passage you chose. Today, a causeway, opened in 1960, carries the road across the gap in an easterly arc over a succession of sea-girt rocky outcrops, with a short midway stride across the western tip of the island of Grimsay.

'There was no causeway in our day – it was a boat,' said Simon. 'And, if it was a rough day in a wee boat going over, you would be soaked! But I never walked the ford.'

'My late father was involved in that, in the 1920s and '30s,' said Ian. 'They arrived with a big drove of cattle and they walked the cattle across and the cattle had to be led by a person that knew the tides . . . or there would be one or two folk with a cow on a rope and they followed the track of the sands, because the sands can be very dangerous and the tides are dangerous. That was all stopped before we got out there.'

'I went across there with the cattle once when the tide was out,' said Neil. 'You had to be quite quick. Well, there's one channel there you had to take your boots off.'

Little did I think that soon I would be taking my boots off and wading that channel myself.

# 6

# *The North Ford*

A photograph from the archive showed a stone jetty with an open boat alongside, well loaded with several over-coated and hatted passengers – some seated, one or two standing. A handwritten caption identified the group as dealers, auctioneers and even a veterinary surgeon from the Department of Agriculture. They were about to leave Benbecula for North Uist. On the jetty, a dark-haired young man held the boat steady with a firm grip on the gunwale – Ewan Nicolson, like his father before him, was the North Ford ferryman. He it was who took Ian and Simon and Neil across the water for the second part of their droving week.

'That's my brother,' said Anne Burgess.

One result of my exhibition at the North Uist show had been an article in the local press and one result of that was a letter from a lady in South Uist who told me that I should contact Anne Burgess in Grimsay. The writer added that Anne's father had been the postmaster and had operated the North Ford ferry. If the tide were in, he would row you across by boat; if it were out then he'd take you over in a pony and trap. This letter had led to an exchange of correspondence, a phone call and an invitation to visit Anne at her home overlooking the ford. She had seen her seventieth birthday come and go and maybe a well-hidden decade more. When I arrived she was busy in the garden and bustled to settle me in her sitting room while she put on the kettle. She was keen to hear all about my travels and what I was trying to do. Tea and biscuits, voice recorder, photographs, pencil and notebook – we were ready.

From the window, Anne pointed out the Grimsay Post Office where she, Ewan and their sister Joan had grown up. Perched on a knoll at the edge of the water, the family home was within sight and sound of the cattle as they approached the ford from the Benbecula sale and began their trek across the sands. Joan lives in that house yet, though it's no longer a post office. The Nicolson children were well used to crossing the sands and Anne regularly rode her pony across from Grimsay to the blacksmith in Benbecula. There used to be horses at the sales too, she said. Although she hated the thought of the calves leaving home to be sold at just one or two years old, she was more concerned about the horses.

'I would plead with my father to buy them all.' Needless to say her pleas were in vain and the horses would be sent away with the cattle.

'Hello, are you in?' A voice from the back door.

'Now you'll get stories,' said Anne and introduced her cousin Angus from along the road. She was right. Short and sturdy, full of life, with a ready laugh, shining eyes, and a head of white hair, Angus had travelled the world, returned home to Grimsay, taken up running and was still competing in marathons in his seventies, despite the after-effects of a broken leg some years before. Like Anne, he had known the North Ford from childhood. They both remembered hearing the bellowing cattle as the drove left the market stance. Then the herd would come into sight as it neared the ford.

'We would look out and the whole Oitir Mhòr* would be full of cattle,' said Anne. 'Maybe two hundred at once and the buyers on horses.'

I had a photograph showing such a crossing: a dealer on horseback, the drovers and dogs urging a large herd of cattle across the sands. Angus remembered going out to one of the small islands in the ford with his pals, where they sat on a big stone cairn to wait for the drovers and their charges to come past. He could show me the place. We could walk the ford.

'Would you like to?' asked Anne, though I'm sure she could already see the answer in my face. 'I have spare wellingtons, size six; you can borrow them.'

No matter that my feet were a whole size and a half bigger than hers, I could get the boots on and that would do fine. These generous people had only just met me but there was no hesitation at all. If the weather held, they would guide me on foot across the sands. There was some discussion about the tides, agreement that Joan should be invited to join us, and that we should use two cars to shuttle everyone to the start and back from the finish. We'd meet at three o'clock the following afternoon.

I drove, dazed and delighted, back to my hideaway in the North Uist dunes. Kettle on, van door open to the sight and sound of the sea and the ringed plovers calling from the edge of the surf, I settled down to type up my notes and recordings. It was difficult to concentrate. I was going to walk the North Ford.

* * *

---

* Oitir Mhòr is the name given to the tidal sands of the North Ford.

We left dry land not far from the old slipway at Gramsdale in Benbecula, where Ewan Nicolson had taken his droving passengers on board. Angus led the way, pointing northwards with his stick in the direction of Carinish, our destination. I tucked my talisman cattle shoe into a pocket. Two hours of walking lay ahead of us. To either side, like stage curtains opened wide, waited the tides of the Atlantic and the Minch.

All we were missing was 200 cattle. I tried to imagine a drove on the hoof: steam rising from the jostling animals, the thud and splash of cloven hooves on hard sand and through remnant pools, the anxious cattle voices, the drovers shouting, dogs barking, the trail of prints and dung that would remain for a few hours before the curtains closed and the sea swept away all sign of their passing.

The old photographs showed drovers carrying dealers piggyback through the water but there was none of that for us. At the first channel, Angus searched for the best crossing. He probed the sand with his stick, waded out to test the depth of water, announced that here we could cross safely. With time and tide always in mind, he allowed no hesitation, urged us on. It was off with the boots, just as Neil had said, and into the current that flowed swift and strong to the Minch. The water reached knee-high and a bit more, flinching cold at first, then bearable enough until we reached the other side and the sun warmed away the goose bumps.

I opted to keep my bare feet, Angus did the same and we walked on. Corrugated sand beneath unshod soles. A skirl of oystercatchers. A dishevelled pile of stones – a cairn, said Angus.

I hadn't expected a cairn. There had been a line of them, he said, to mark the safe route across the ford. Taller than a man.

Some kept their heads above water even at high tide. Most have disappeared, broken by storms and scattered by the ever-meddling sea. Others are hardly distinguishable heaps, though a keen eye can still detect traces of the skill that made them. Every year the sea dislodges one more stone, widens one more crack. For now, enough remains for those who know to show those who don't how it used to be.

I thought of the people who came out here on the sand to build the cairns, by hand, stone by stone, in the intervals between tides – people who were used to working with stone for their survival, accustomed to building harbours to shelter their boats and houses to shelter their families. What was it like to bring cattle across the ford, following this lifeline created by your fellow islanders? What if you yourself had been one of the cairn builders? Did you, in passing, place a knowing hand on stones you had lifted and set in place, intimate with this cairn as with the walls of your own home? Angus must have felt some echo of this down the generations – he revealed that he had restored one of the cairns to its full height, reluctant to let all trace of the old way disappear. He would show me, he said. We walked on.

The sea was nowhere to be seen. How easy to be deceived into thinking we had all the time in the world. How easy to stray from the safe route into dangerous soft places that wouldn't bear the weight of a cattle beast. Angus knew those places. Some had shifted with the building of the causeway, he said, and I suddenly felt insecure. When my feet sank an inch or two in an isolated patch of softish mud, I pulled them out with a mild thrill of panic. Thousands of years ago all this was dry land, Angus told me. Then the Atlantic broke through the western sandbanks, came up against the waters of the Minch flooding in the opposite direction, and the two have been meeting and parting ever since. He spoke of this geological yesterday as though it were a living memory. Such is an islander's attachment to his native environment.

We stopped. We were halfway across the ford. The horizon that I had mentally made my goal proved to be just one more deception – a short, broken chain of small islands reaching out into the sands, left dry and a few feet high by the absent tide. Angus was pointing. Look, there on the grassy top of Eilean na h-Airidh was the cairn he had repaired. This was where he and his friends had come, running across the narrow tidal channel from Grimsay to watch the cattle pass. And there, pointing again, the gap between the two nearest islands – the crofters had shifted boulders to one side and the other by hand, leaving a level passage for carts and cattle, horses, dogs and men. We walked through. Seaweed and small stones. My feet slipped and winced after the kinder surface of the sand. I imagined hundreds of cattle feet doing the same. The boys would be sitting on the cairn as the animals crowded into the gap. Again the dogs barking, the cattle complaining, the drovers shouting and no doubt finding time to greet the youngsters and field a stream of excited questions: 'How many today?' 'Were the prices good?' 'Is that a new dog you've got?' And on they'd go, from cairn to cairn, aware of the tide.

The fog could roll in without warning, said Angus, and night could come on you, or you might be travelling in the early hours to catch the tide for the sales and then you couldn't see from one cairn to the next. It was easy to become disoriented. So, in places, there was a line of stones set in the sand like stepping stones between the cairns. You could follow these. Even more vulnerable than the cairns, almost all have sunk into the sand or been shifted out of line. Angus showed me a remnant – one, two, three, four, each with its crown of seaweed – and I wondered how it felt to be not-quite-lost out here in darkness or swirling mist, trusting this thread of stones to lead you to the next cairn, listening for the sound of creeping water.

As we emerged from the cattle gap, we saw the rest of our party ahead of us. Having baulked at the earlier boots-off, over-the-knee paddle, they had leap-frogged our progress by driving round to the causeway and rejoining the route at Eilean na h-Airidh. With that island and its companions at our back, a further small-island gathering on our right and the coastline of North Uist ahead, this second half of our trek felt less exposed, the sea less imminent. Another deception, of course. The Atlantic, when it came, would push – no doubt with some force and speed – into this ragged arena along a channel much deeper than that which had deterred our booted companions. Meanwhile the Minch would be sneaking in from the east, fingering its way round Grimsay's fragmented northern edge. The drovers would be wary of any dawdling and so was Angus. We chatted our way steadily towards Carinish, wearying a little now, splashing through the shallow delta-like branches of that main incoming channel.

And so we entered the Bàgh Mòr – the big bay – wide of mouth with a throat full of stinking, glutinous, hungry mud. After two hours of clean sand, my feet were black to the ankles in minutes. No matter – I had walked the North Ford.

Like the drovers, we had left transient evidence of our journey – our footprints and the pock marks of Angus's stick spelling out progress, diversions, pauses for discussion. These would bear witness for a few hours then all trace would be swept away. In years to come, the old stones would join them. Then, save for one restored cairn and the memories of a dwindling generation, the route of the North Ford crossing would be no more. And yet . . .

Several months later and many miles from the Western Isles, I unrolled one of the old maps that Ian Munro had insisted I should find: Sheet 23, Crown Copyright 1959, Scale One Inch to One Statute Mile. No causeway. Two roads stopped, like broken string ends, to north and south of the space named Oitir Mhòr between Benbecula and North Uist. Across that space, joining the strings, a double dotted line followed exactly the track of the cattle droves, and that of my bare feet. In the days when Ewan Nicolson ferried Ian, Neil and Simon from Gramsdale to Carinish, the North Ford crossing had been an officially recognised route, preserved here in print by the Ordnance Survey. There was more. Further dotted lines wound among the skerries and islets: Anne and her horse would have followed this one from Grimsay to Benbecula; along here her crofter neighbours would have driven their beasts to the sale in North Uist. Each of these trails held a story and I had been privileged to meet people willing to share them with me.

# 7

# *Six Down and Two to Go*

The biggest sale of the week was yet to come. A few miles north of the drovers' landing place in North Uist, where the main highway to Lochmaddy met the island's circular coast road, was Clachan.

'It consisted of just maybe four or five houses, a shop, a post office and that was it,' said Ian. 'But it was the heart of that area. And that was a big sale – there would be maybe two hundred to three hundred cattle in the one day.'

'Start by eleven o'clock in the morning,' said Simon, 'going to six o'clock, non-stop. Big sale.'

On a small hill behind the Clachan Stores, I found some wooden pens. Was this where the market had once been held? I found the answer in a croft house by the sea at Kyles Paible, west along the coast from Clachan, where Ena Macdonald and her son Angus breed pedigree Highland cattle. In the 1920s and '30s, when Ian Munro's father came out to the islands, almost all the island cattle were Highlanders but, by the time Ian himself arrived in the 1950s, preferences were changing.

'Maybe about fifty per cent would be pure Highland when we first went out,' he said, 'then there would be more Shorthorn and Hereford. Pure Highlanders in the 1960s were the minority.'

When Ena took over the croft from her father, she decided to stick with a purebred herd and now the Macdonalds' Ardbhan fold is highly respected worldwide.

'We got the first pedigree bull in 1985, I think,' she said. 'It was in 1988 I sold the first pedigree bull. We've sent them abroad, all over the place.' Ena was now in her seventies and, while she was happy to have Angus carry much of the workload, taking a back seat was not part of the plan.

'I was born here,' she said. 'I left home for school in Inverness when I was fourteen. I just stayed there a year and then I worked with my father for a year and then I sat the Civil Service exam and I went to Glasgow. I was there for nearly five years and I got married in December 1960. Then we went to Australia the following February. I came back from Australia in 1964, Angus was born in 1965, then I went back again but I was so homesick. My father was in his sixties so I just worked here on the croft. My father died in 1975. You could say that from 1966, I've put my heart into the soil.'

Yes, it was good soil, she agreed, but 'it's the wind. We get terrible winds out here. So constant, you know, in the winter. Some years are worse than others but it's quite hard.'

Ena confirmed my guess about the location of the Clachan sale. 'It hasn't changed that much really,' she said. 'And the house opposite, that's where old Susan used to stay, and on sale day she used to sell teas. Everybody went in there and got tea and sandwiches. And I've got the mantelpiece that was in that house – come on and I'll show it to you.'

We went through to the sitting room and there, above the fireplace, was a thick slab of wood complete with burn marks where lighted cigarettes had rested over all the years of the cattle sales. What stories it could tell!

'There's another lady,' said Ena, 'a cousin who lives across the water here. She's eighty-nine. She and her husband owned the Westford Inn. I know that, after the sales at Clachan, people used to call there and she would make meals. They didn't sell meals but on sale day they were always feeding anybody that called. In fact, I think some of the drovers stayed there.'

Cattle would converge on the Clachan sale from all the southern townships of North Uist: from Balranald in the west, Locheport in the east, and from Grimsay across the ford. It was 'a huge sale', said Ewen Macleod, who watched them coming while waiting with his paint tins for the sale to start.

'There's a wee hill they used to come over,' he said, 'and I thought surely that's them all now. Then another half a dozen folk would appear, another twenty. They herded them in lots. It was lovely, you know, to see them.'

'We used to watch the cattle walk to Clachan,' said Ena. 'Quite a few hundred of them – it was quite a sight. Today with all the traffic on the road, oh my goodness, it would be just impossible.' She remembered taking their own cattle to the sale too. 'The tide would be out. We would take them across the strand here and we'd meet up with others and walk from there to Clachan. I can remember – see, in these days, nobody sold calves like they do now. The calves were left until they were a year and a half. They'd be born say April, May. You'd be selling them a year next September, nearly always the September sale. I can remember one time, we had a wee heifer calf and we were selling it. I used to make such a fuss of them, you know. It was a real pet. Anyway we got over to Cladach Eileanaich, near Clachan, and there were some cattle beside the road and there was a cow the same as our Rosie, on a tether nearby. And, when the calf saw it, she thought it was her mother and she went running. Oh, you know something, I just could have cried. I never forgot that. I used to feel it for the cattle, you know.'

Once the sale started there was no pause until the last bid had been made, the last paint-mark and scissor-clip administered, the last beast led out of the ring. A long day for everyone but an even longer one for the drovers and their dogs.

'When that sale was completed,' said Ian, 'that cattle had to be walked to Lochmaddy.' Following the usual pattern, there would be diversions after runaways who thought of heading for home then, once the drove settled down, the twelve or thirteen miles to Lochmaddy would be covered at a steady plod until they reached the outskirts of the village. A crowd of excited schoolchildren would come rushing along the road to meet them, until 'you weren't sure if you were herding cattle or children', said Ian.

'It was maybe getting dark before you would reach Lochmaddy,' said Simon. 'You would get some grub at Lochmaddy Hotel or somewhere like that.'

* * *

One more sale for the drovers and one more visit for me. Ian was very keen that I should go to see Alasdair Macdonald, known as 'Ahmore' after his home, site of the eighth and final Uist sale. I had seen Alasdair at the North Uist show – a tall, distinguished, elderly man constantly immersed in conversation or intent on watching the judging of cattle and sheep. I didn't get a chance to introduce myself that day but, when I phoned towards the end of my Uist travels, he was delighted to hear news of Ian Munro and said, yes, I must come to see him. He'd be around the following day. I arrived to find no sign of Alasdair – oh, he was out moving cattle, said his daughter. He'd be back soon. He had ostensibly handed over the reins to his sons but there was no way he was going to sit back and put his feet up. Like Ena with her Highlanders, Alasdair had farming and cattle in the blood. Retirement was an alien concept. When he returned, I was greeted with a huge handshake. He settled into a chair and asked me about myself. This kindly gentleman was shy and courteously wary. We shared photographs, Alasdair pointed out characters he had known and gradually our chat teased out wisps of stories that added bright spots of detail to the bigger picture my island journey was building.

'There was a man from Perthshire,' he said. 'Balquhidder – I think he was coming here in the 1960s and '70s. He used to buy cross-Highland heifers and they were doing very, very well for him. But he was saying that, the first year, he'd have to give them a lot of salt licks. I suppose from being near the shore they were used to a lot of salt. But after the first year they were alright and he was very pleased with the way they were doing for him, the island cattle.'

Had Alasdair always lived here? Well, almost.

'The only move I did? Across the burn! The old house was on the other side.' The school was about three miles away and he walked to classes every day. 'No word of school transport then or school dinners either. Bottle of milk and a sandwich, that was it.' What of the cattle sales? They were on his father's land so, yes, he'd been involved all his life really. And did he skip school on sale days? 'Oh well, we tried to get off anyway and, ach yes, you'd be running after cattle here and there, helping. My mother used to serve teas in the house at the sales and my wife used to do it after that. It was a busy day for them.

'My father came from the other side of Uist. He wasn't from this side originally. He came here in 1919 after the First World War. The whole of Ahmore was a farm when my father came here. He was paying £25 a year. Well, he would need to sell four or five stirks to make that money and there's no word of any subsidy or anything at that time, you know. In 1924, the farm was broken up into crofts.'

As children grew into teenagers they were given more responsibility. Alasdair's father and uncle used to buy stock at the Clachan sale and the youngsters would be expected to get the new cattle home – on foot – with the added bonus of some entertainment in the evening.

'They used to put our beasts along with the mainland cattle,' said Alasdair, 'and we would walk them to grazing just this side of Lochmaddy, at Sponish. They'd be held there for the night and the auctioneers would send the hotel taxi for us, take us down to the hotel and get our dinner for helping them. Very nice of them indeed and, after that, there'd be a dance in the village hall. We went to the dance. After the dance, two of us would go to Sponish, sort out our cattle and start walking them home and we'd be in Ahmore before the sale started! We were young, we never thought anything of it.'

Early on Thursday morning, Alasdair would be almost home with the drove, while Ian, Neil, Simon and their dogs were in the bus with the rest of the auctioneering team and the buyers, heading for the last sale of the week.

'Ahmore,' said Simon. 'That place was worse than the other places!' The sale would finish in the early afternoon and the cattle were walked into Lochmaddy to meet the boats. 'Maybe you would be walking and, if two beasts at the same time would burst out, the dog would only manage one of them. He would keep it and the other one was away the other way and you were trying to get at him and, oh, gee whizz. Maybe sometimes miles before you would get them. Sometimes you would be on the road for two and three hours – there would be only maybe six cars behind you. Two or three minutes today and there's twenty!'

'Och, you couldn't go on the roads today with cattle,' said Neil. 'We were lucky enough. There wasn't much traffic on the roads, no. The odd car . . .'

'Oh and trying to get them past,' said Simon, 'and the cattle would burst this way and that way and maybe some of them had never seen a car.'

'Well, it would be totally, totally impossible to walk them to the market today with the traffic and single track roads and – oh, it would just be impossible,' said Alasdair.

At the Lochmaddy pier the sorting and loading were as troublesome as they had been at Loch Skiport and Lochboisdale.

'No pens on the pier or anything,' said Ian. 'You just made do with gates and that.' They had to round up cattle from the promontory next to the pier and drive them down to the boats. Orange paint marks for Oban, green for Kyle. Drovers, auctioneers, buyers – everyone lending a hand and, no doubt, some of the local youngsters would be eager to join in, the drovers shouting them back out of the way for fear they might get hurt.

As the boats headed down Loch nam Madadh and out into the Minch, island life would return to normal until sale week came round again in six months' time. For the local drovers and their dogs, there would be a chance to relax briefly before getting back to their crofts.

'When the sales were finished, the dogs were just lying down beside you,' said Simon. 'There was no' much left of the dogs' paws.'

'At the end of the sales, they couldn't walk,' said Neil.

# 8

# *Cattle Boats and Iron Rails*

'They would take the drove to the pier head and then they were all loaded,' said Kenny McKenzie. 'They would go on to a cattle boat up a narrow ramp, only about a metre wide. The boys in the boat, they knew what they were doing and they would put them into pens. Five or six or eight cattle in a pen. Then they would come off in Kyle and they would go on to the railway wagons.'

The drovers were on their way home but, for the cattle, the journey was far from over and they would never see home again. They would be exhausted and footsore by the time they reached their final destination, which could be somewhere in Aberdeenshire or Easter Ross, maybe Stirlingshire or even across the border into England. It was a far cry from the machair of the Western Isles.

The drovers travelled with the cattle, Ewen Macleod remembered. 'The crew would keep an eye on them too,' he said, 'and the crew would come to you if there was a beast down or something. There must have been a lot of pneumonia. You couldn't see them for the steam, especially if they went on wet. Some of them were just walking on pins, you know. The heat in their feet – walking on the roads then going on to the hard steel on the boat. I've seen cattle and they could hardly stand, with the sore feet. Dealers would say it would take a fortnight when you got them home, before you would say they were sound.'

In the early days of Ian Munro's work in the islands, the boat for Kyle of Lochalsh would normally leave Lochmaddy, head round the north end of Skye and down the east side of the island, the route and the timing depending, as always, on weather and tides. Occasionally, said Ian, the homeward voyage would take them south through the Minch, east between Canna and Soay, round Point of Sleat and then north up the Sound of Sleat, negotiating the turbulent straits at Kyle Rhea, where tidal ebb and flow surge through the narrowest point between Skye and the mainland like a river in spate, complete with whirlpools and standing waves. I once watched a Royal Navy vessel chancing its arm here against the current and losing the contest, first brought to a standstill, then pushed slowly but surely backwards to wait and think and try again. In the 17th and 18th centuries, long before the auctioneers and the railways, this was the main route for cattle from Skye to the trysts at Crieff and Falkirk. Drovers brought the beasts down through Glen Arroch and swam them across the channel. It was a skilled operation, with an annual passage of between 5,000 and 8,000 head of cattle.

> They are forced to swim over Caol Rea [Kyle Rhea]. For this purpose, the drovers purchase ropes which are cut at the length of three feet having a noose at one end. This noose is put round the under jaw of every cow, taking care to have the tongue free. The reason given for leaving the tongue loose is that the animal may be able to keep the salt water from going down its throat in such a quantity as to fill all the

> cavities in the body which would prevent the action of the lungs; for every beast is found dead and said to be drowned at the landing place to which this mark of attention has not been paid. Whenever the noose is put under the jaw, all the beasts destined to be ferried together are led by the ferryman into the water until they are afloat, which puts an end to their resistance. Then every cow is tied to the tail of the cow before until a string of 6 or 8 be joined. A man in the stern of the boat holds the rope of the foremost cow. The rowers then ply their oars immediately. During the time of high water or soon before or after full tide is the most favourable passage because the current is then least violent. The ferrymen are so dexterous that very few beasts are lost.*

Those men knew the wisdom of patience. A current that can bring several hundred tons of military steelwork to a halt is not to be challenged by men in rowing boats towing cattle.

The arrival of the cattle boats in Kyle would be quite an occasion. Norman Finlayson and Ewan MacRae grew up in the village. Both watched with interest the transfer of cattle from boat to train, although neither was directly involved.

'I was born in Balmacara,' said Ewan, 'then we moved to Kyle. I went to Glasgow, did my apprenticeship, then I came back here in 1962. With a guy from Berneray, Harris, we set up an engineering business on the pier – fishing boats, yachts and CalMac boats. They had all these launches. They were all Board of Trade registered because they were taking passengers. This was before they put the piers into the likes of Eigg, Rum and Canna and the launches used to go alongside the steamer and offload [passengers and goods] back and forth. But they were all kitted out for cattle because right round above the

* James Robertson, *General View of the Agriculture of the County of Inverness*, 1813.

seats where the passengers sat, there were rings for every beast. When they put the cattle out to the boat, they lifted them out of the launch on board.'

'My father was the policeman in Kyle here,' said Norman. 'I've been about boats since I was just out of nappies, I would think. We were always on the pier as kids. Oh it was busy in these days, Kyle. My first strongish memory would have been about 1960. That would have been the *Loch Seaforth* that used to take the cattle. She always berthed on the west side of the railway pier. There was a door on the side [of the pier] where the cattle would come up this concrete ramp, sliding all over the place. There were actually two ramps on the west side of the pier and one on the face of the pier. That one's still there, right opposite the station. It was used more for the *Loch Dunvegan* but the one that the *Loch Seaforth* used has been bricked up now. The *Loch Seaforth* took the cattle from Lewis and the *Loch Dunvegan* did special runs to the Uists for cattle.'

I found the remaining ramp and ventured down the steep incline. Stone ribs, slightly raised and built crosswise into the slope, would have given the animals some rudimentary assistance as they struggled up into the daylight. At the bottom, a heavy metal grille denied inquisitive visitors access to the water. It was a dank, dark place with a picture-postcard view of Castle Moil on the Skye shore, framed in the opening through which the island cattle would have first set foot on the Scottish mainland. Ian Munro had described for me how a door in the side of the boat would be lined up with the hole in the side of the pier and the cattle driven through. It would hardly be a reassuring disembarkment for terrified beasts after a rough sea journey lasting several hours.

'It was always a case of shouting and bawling and whistling

and sticks,' said Norman. 'The men had a terrible job controlling them – running them along the west side of the pier to the cattle pens which were just below the fuel tank that's there today.'

'You can still see the area where the pens were –' said Ewan, 'where the new doctors' surgery is, underneath the railway bridge, an area that was cut out of the rock. The big complaint was the noise they were making because they were right below the Kyle Hotel. They used to keep everybody awake with their bawling through the night!'

Everyone would be keen for the cattle to depart, especially if their transport had been delayed. Many of the trains on the Kyle line were 'mixed', which meant they had passenger carriages as well as a goods wagon or two but, when the island cattle were coming in, the railway company put on a dedicated cattle train.

'There was always one or two [cattle trucks] left here in Kyle,' said Ewan MacRae, 'but, when they were moving a boatload, they would take in maybe say ten trucks.'

Things didn't always go according to plan, as George Tait remembered. 'We were needing wagons at Kyle and they couldn't find any,' he said. 'Well, they didn't know where they were. That would have been probably the mid 1960s. The livestock wagons just weren't available.' No doubt the problem was eventually resolved and there would be more sticks and shouting and bawling below the Kyle Hotel as the pens emptied and the cattle wagons filled.

'I don't know how many beasts in each wagon,' said Ewan MacRae. 'Probably they'd be packed in that tight, they couldn't fall over.'

Kenny McKenzie remembered that it was 'twelve to fourteen in a wagon'.

Away they'd go, coast to coast. Even today, it's a sedate two-hour daunder with time to admire the scenery and count off the litany of station names along one of the world's best-loved railway lines: first the coastal villages of Duirinish, Plockton, Duncraig, Stromeferry and Attadale to Strathcarron; then the steep climb out of the strath and up Glen Carron, with maybe a pause for breath at Achnashellach before the final push to the watershed.

The children of Achnasheen – halfway point in the rail journey – would hear them coming as the train began the long, gradual descent into Strath Bran with its wide river flats and herds of red deer. Before long the sinuous line of cattle wagons would come into view, hauled by an engine puffing clouds of steam, each truck crammed with anxious beasts and trailing its own vaporous banner of sweat and cattle breath. Would the deer, catching the scent of animal distress on the disturbed air, trot a few unsettled strides to a safer distance? Would the children be distracted from their lessons?

'Here they come,' someone might shout. 'The cattle are coming!' and there would be a rush from the schoolroom down to the station to wave at the engine driver, who would wave back. At close quarters, there would be hands over ears for the hissing engine and the bawling cattle. Small fingers would hold small noses against the pungent smell of frightened animals. Maybe a little girl would gasp on spotting the flared nostrils and bulging eyes of a two- or three-year-old stirk that had – all in a week – been taken from his mother, pushed into a circle of strangers who prodded him with sticks, daubed with paint and clipped with scissors, then chivvied with dogs along the long walk to Lochmaddy where a bid for freedom would end in more shouting, more prodding. He had endured the stinking, slithering, pitching and tossing of a sea crossing. Emerging into a further bewilderment of strange smells, alien voices and more barking dogs, he had been crowded into this jolting box. The peace of his calfhood had vanished forever.

As the train pulled out of Achnasheen, the pupils would return to their studies, full of excitement that they had seen the island cattle on their way to the big town in the east. The cattle travelled on, through Achanalt, Lochluichart and Garve, up by Raven's Rock and down to the River Peffer, over three clamorous level crossings and round the heart of Dingwall town to the station. They had arrived.

'The mart had fields just adjacent to the station,' said Kenny McKenzie, 'so the cattle would be let out into those fields.' Fresh air, solid earth under sore feet, plentiful grazing, and freedom to move. The fertile alluvial soils beside the estuary of the River Conon would certainly provide a good feed, and the overnight rest would be more than welcome. The relief was to be short-lived.

# 9

# *Hooves Along the High Street*

'On the Saturday, the yardsmen in Dingwall would walk all the cattle up to the mart,' Kenny said. 'There would have been maybe two or three hundred being walked up. At five thirty in the morning, we had to gather [the cattle from] the fields and walk them through the streets. We had to try and get them through before six-thirty or seven in the morning. Men with dogs gathered them up and we then placed men through the town to watch all the streets. So we actually drove [livestock] through the streets in Dingwall, very much to the disgust of the shopkeepers. Their customers were walking into the shops with that funny stuff stuck to their feet and they didn't like it. Never mind, they had to put up with these things.

'Then they would all be sorted. You would let one run along a passageway and you would just shout, you know, "Left hook!" and the next one would be "Right hook!" and the next one would be "Left shoulder!" and all this. And you would pen them up and then you'd go on. If the paint marks weren't there, the scissor clips were there. Then you'd count them and say, "That's twenty-five belongs to so-and-so – that's his mark." There was a lot of work in it.'

Ian Munro remembered that 'most of the stock that went

away from the mart, if they didn't go locally, they were put on the train. A lot of the buyers came from Aberdeen, Inverness, Elgin.'

The early-morning excitement would be over by the time young Beatrice (Trixie) Mackenzie arrived at work, though she needed to watch where she put her feet. 'There's seventy years since I started work as a young girl in Dingwall,' she said. 'It was a draper's shop [on the main street]. In the afternoon when the cattle would be going down to the station, some of them would come tearing down the street. Sometimes you would hear them coming, you know; you would hear this man shouting at the dog. He looked sort of old to me but perhaps he wasn't that old – I was just a teenager. I can remember other younger ones too. Of course, you see, the year was 1942 when I started working there. And, you know, that was the time, the war and that.'

With men of fighting age away from home, the older and younger generations would have been essential members of the team. As life returned to normal after the war, two young schoolboys couldn't resist the lure of the mart: Ian Munro and his friend George McCallum

'After the school bell went, we went down,' said Ian. 'Our job was taking the stock to Dingwall station, and they'd walk right down the street.'

'Some would just come down very placid you know,' said Trixie. 'They would saunter down past the shop.'

Some, perhaps, but not all. Broken windows were a regular hazard. George Tait remembered 'going to Dingwall railway station, unloading wagons and walking the cattle up the street and it was always terrifying, wondering how many windows we would break.'

'There was a lot of glass about the shops then,' said Trixie, 'and it wasn't like the double glazing today. Oh many a clatter. I can remember hearing this heavy bang and I said, "What on earth is that?" And this was Hepworth's, a gents' shop. It had more glass than the usual ones.'

Ewen Macleod recalled similar scenes when cattle arrived off the boats in Oban. 'Most of them came in to the railway pier and would come up from the pier and they would turn right a wee bitty, over the bridge. It's just thunder, the way they went. If a beast would see its shadow or sometimes one passed a shop window, you have three hundred cattle coming to a sudden stop. The next morning, there's a queue of folk at the mart office: "You've broken my mirrors on the car!" "You've broken this!" "You've broken that!"'

Trixie had an even more adventurous experience. She and a colleague were at work in the draper's shop when the cattle came by on their way to the station. Stress levels were no doubt high among the beasts after a morning of being pushed, prodded and penned in yet another strange place, then finding themselves once again herded through the hustle and bustle of a town. Uist and the machair were a long way away.

'You know how the shops were then,' said Trixie. 'There was just the counters. The other lady that was with me, she was awfully frightened of the cows. If she would hear them she would run through to the other end of the shop. It didn't used to bother me because I was brought up in the country. Oh, quite a lot of them would look in. But, look, this one came right in and – I don't know why, I don't think I was frightened or anything, but I was fit then – I jumped over the counter and slowly it went back out. No it didn't do any damage but it was quite a big cow. You know, it just came in just as if [to say] how are you today. Oh, poor Bessie was absolutely paralysed with fright.'

Once a week, a sale of livestock took place in the mart. Situated as it was right in the centre of town along with the slaughterhouse, the mart had no small effect on everyday life for the townsfolk. Trixie's niece, Sandra Bain, was a schoolgirl in Dingwall during the 1950s. She remembered how the weekly sales affected her after-school journey home to the Black Isle village of Tore. 'Wednesday night being market night, the bus was half an hour late leaving, so we would be stravaiging the streets. It was a service bus that ran between Kessock and Dingwall – Archie's bus. He did a run to Tore at four o'clock on a Wednesday and then another at five. (It was half past four every other day.) So we got the five o'clock one from school as the four o'clock one was too early. The bus was late because it was sale day – the men were at the sale and the women were in doing their shopping.'

'That was a busy day for the town,' said Trixie. 'They would all be coming in.'

Livestock would be coming in too, from the farms round about and from further afield. By the time Kenny McKenzie started his career with the mart in 1965, 'trains still had livestock wagons but that was mainly for stock from the west coast, from Uist. The bulk of the local stuff from, say, Easter Ross, would have come in by lorry – not such big lorries [as today]. If you went back ten years before that, there would have been stock walking into the town.'

# 10

# *All Alone on the Dirie Mor*

If Duncan Stewart had joined the army, he might never have found himself walking from Ullapool to Dingwall with a herd of cattle belonging to the most famous drover-dealer in the north of Scotland. Europe was embroiled in the Second World War and young Duncan had been turned down by the recruiting office, even though he had passed the medical inspection with flying colours. He was a shepherd and the army didn't take shepherds. They were needed at home to help feed the country in these difficult times.

Born into a crofting family in Achiltibuie in 1924, Duncan was 14 years old when he left school. His mother told him he'd have to leave home and make his own way in the world – their ten-acre croft couldn't support a family of six. It didn't take long to find work. Captain Lawson, owner of the nearby Inverpolly Estate, needed someone to take care of his dogs and a small power station – a curious combination that worked well for Duncan, although the pay was minimal.

'Eight pounds for three months,' he said, 'but I got my food and it was a very good job for a boy.'

He walked Captain Lawson's dogs for two years, then moved to the Rhidorroch estate as caretaker of a shooting lodge. 'That's when "The Drover" got a hold of me,' he said.

The Drover was Kenny Macrae of Ullapool, a dealer well known in the Uists. He came from a family of cattlemen: his father, Roderick, was a drover; his grandfather, Donald, was a drover; his uncles, Hector and Don, were both drovers. Their young relation took up the family calling with enthusiasm, travelling all over the north Highlands and the Western Isles, buying cattle and sheep and often horses as well. Kenny was an astute businessman who thought and acted on a large scale. He would return from his island trips with four or five hundred cattle, loaded on to a boat in Stornoway and thrown overboard several hours later, close to the Ullapool shore. The enforced swim would help wash off the effects of a long, crowded and often stormy crossing of the Minch. It must have been quite a sight, as the cattle lurched out of the tide – no doubt an irresistible draw for many a crowd of sightseers and youngsters sneaking away from the classroom. The young beasts would be kept until they were ready for market the following year.

All that livestock needed to be fed and, during the war, Kenny rented the vast grazings of Rhidorroch, along with grazing rights over several other estates in the area. As well as cattle, he put thousands of sheep on the land and, when the war ended, the sheep were walked through the hills to the Lairg sales, joining other great flocks from all over the north Highlands.

The Macrae enterprise didn't stop there. As well as a butcher's shop in Ullapool (where black pudding and haggis were made to a secret recipe), Kenny had a slaughterhouse on his farm at Auchendrean, which lay to the south of the village. Auchendrean was to be Duncan Stewart's next home.

'He says you come and work for me now – work on the farm,' said Duncan and soon the dog-walker-turned-caretaker was immersed in his new life of farming and shepherding at Auchendrean. 'I enjoyed every minute of it,' he said. 'Oh, they were good days – hard graft but good days. There were only maybe a hundred and fifty acres to farm, but then there was an outrun of thousands of acres, all stocked with sheep and cattle.'

At the approach of autumn, Kenny The Drover prepared to head west across the Minch in search of fresh stock. He would be away for a week or so, a well-kent figure among the other dealers as they followed the auctioneers round the market stances in the islands. He would return to the mainland in time to sell his own cattle at the Dingwall mart and was confident that young Duncan Stewart could walk them safely across fifty miles of exposed country on his own.

'I was a week on the road there, just me,' said Duncan. 'I was advised to graze them along. Once you got them walking, they understood what you wanted to do with them, especially when you had plenty time and you weren't rushing them. It was quite good weather too, but I still had a raincoat in case.' After all, it was October in the Highlands of Scotland and the route from Ullapool to Dingwall can be wild. Flanked on either side by rugged mountains, the broad upland strath known as the Dirie Mor is wet underfoot and very often wet overhead too, when wind-driven rain, hail or snow sweeps through. Duncan, with his just-in-case raincoat had every chance of encountering stormy weather, yet he was equally likely to enjoy the Dirie in a playful mood of blue skies and galloping clouds, with perhaps a touch of autumn mist about the mountain tops, maybe the echo of a late-rutting stag in the high corries, an eagle shadow passing across the hillside. Nothing escapes an eagle eye and the herd's steady progress through the glen would certainly be noted from afar. A subtle shift of wing and tail feathers and a long, angled glide down the air would bring the bird over for a detailed inspection of the intruders. In such circumstances, a young man might think this the finest place in the world. Duncan certainly did.

'It was lovely,' he said. 'I still remember it, very clear in my mind – all alone on the Dirie Mor. It was quite an adventure – oh, indeed, yes!' And his eyes twinkled at the memory. There was precious little shelter along the most exposed section of the journey, but 'I wasn't particular where I stopped, to tell the truth,' he said. 'I was quite happy just to be with the cattle.'

In those days of the 1930s and '40s, Duncan and his beasts really did have the Dirie Mor to themselves. There were no fences, no traffic, no tarmacadam – just a thread of gravel tapering away to the horizon and the cattle grazing their leisurely way along. Walking cattle along that road today would be perilous. As the main trunk road between Inverness and the ferry to Stornoway, it's wide, fast and busy. Twenty-first-century cattle hurtle by in articulated lorries, travelling from field to mart in a few short hours and cared for by a different breed of drover altogether.

The gradual descent from the highest point of the Dirie Mor at 279 metres leads to Aultguish and the River Black Water's winding course round the foot of Ben Wyvis towards Garve. The new road skims the edge of Loch Glascarnoch, which is much deeper, wider and longer than it would have been when Duncan made his way down the glen. The massive push to develop hydro power in the 1950s and '60s created a dam and submerged the old road. The Aultguish Inn survived the drowning.

Here, Duncan paused for two nights, letting the cattle graze 'on a nice green spot called the Cearcan' on the slopes of Sidhean nan Cearc, a low hill to the south of Aultguish and still marked on the map today. This same spot had probably been used by many generations of drovers, with the inn a welcome haven after the rigours of the Dirie Mor. It was also at the junction of two alternative routes to the Ross-shire markets. One, the old Fish Road (so-called because of the women who carried fish from Ullapool to sell in the east-coast towns), led over the hill by the deer forest of Corriemoillie and on to Garve. Drovers' cattle may well have shared the path with the fish and it can still be walked by anyone willing to tackle the mix of bog-ridden path, open moorland, deciduous woodland and forestry tracks. The second route, longer but flatter, is followed by the modern road along the Black Water. Kenny The Drover had instructed Duncan to take the second option but to stop and wait at Inchbae Lodge under the slopes of Ben Wyvis. This was another comfortable overnight stop for Duncan, with grazing on the hillside for the cattle. Kenny would join them for the last part of the journey. He had in mind a third route that would avoid the road altogether and take them to within a few miles of Dingwall in a single day's walk.

'Next day was the Sabbath day,' said Duncan. 'Kenny was going to catch up with me that day to walk over the hill, between the two hillocks. Monday morning was the sale day.' The 'hillocks' were Ben Wyvis and Little Wyvis and the route between them climbs steeply from the Black Water into the wet glen of the Bealach Mor, then crosses a rough, peat-haggy watershed before descending to Fodderty on the northern slopes of Strath Peffer.

'It was a short cut –' said Duncan, 'an old route, a right of way.'

With permission from the farmer, they stayed at Fodderty overnight and, early the next morning, a short walk along the strath took them into the centre of Dingwall, where Kenny sold his beasts. Duncan's droving journey was complete.

More than 50 years after his passing, mention of Kenny The Drover still brings a light to the eyes of sheep and cattle farmers throughout the Highlands and Islands. His death in 1959 was like the loss of a front tooth – the gap an insistent reminder of his place in Scotland's livestock heritage. The young man he once trusted to walk his cattle over the Dirie Mor will never forget him.

'He was a fine man to work for,' said Duncan. 'He was a wonderful man. They're all gone now, himself and his family – the old droving days, all gone.'

Duncan spent the rest of his working life in farming, shepherding for many years in the west before moving with his wife to the other side of the country not far from Dingwall. I met him at the Black Isle farm where he had lived for over 50 years with his family. Now widowed but surrounded by children and grandchildren, he looked back on his days as an Easter Ross farmer. There had been cattle and sheep and sometimes grain, he said.

'I was a dealer, going to Dingwall every week, buying and selling – you had to be a dealer to survive. Things were cheap, mind you, when I came here, compared with the way the world is now. Oh, I enjoyed it, I enjoyed the life.'

# 11

## *Going, going . . .*

During the week of the island sales, the two companies – Reith and Anderson of Dingwall and Corson of Oban – worked together as 'joint auctioneers'. There had not always been such co-operation between them.

'Before the war,' said Kenny, 'Reith and Andersons would say, "We're going out the second week in April." Then next thing you would see advertised Corson's would be going out the first week, to get out before them. Even then, it was getting more difficult to get dealers to go away from home for a week to buy cattle – so they went together. I think they decided to do joint financing – Reith and Andersons would finance the spring sale and Corson's would finance the back end sale. It wasn't just quite so good because poor Reith and Andersons, all these locals buying, we were having to finance them and then hoping that they would get their money back in the back end.'

This was quite common practice throughout Scotland, and auctioneers helped support many livestock farmers and dealers. But fortunes could be made or unravelled in the sale ring. An island crofter might buy a couple of stirks on credit at the spring sales, put them to graze and grow on the machair over the summer and hope to make a profit by selling them on in the autumn. Hopes realised, the crofter would pay off his debt to the mart and pocket the surplus. If his optimism proved to have been misplaced, he and the mart would both lose out and a dealer from the mainland would get himself a bargain. Sellers, buyers and the mart companies themselves, all depended on the auctioneer's ability to coax money out of well-guarded pockets. It was a skill learned over time, starting outwith the sale ring.

'The first year or two, you'd be doing the accounts,' said Kenny. 'I can remember sitting in a shed, clerking. A lot was done by hand – no computers in these days.' Sheets of paper would come from the sale ring with details of who had bought what and for how much and the clerk would need to check them. 'You'd go back and forth to the sale, collect the sheets yourself so you could see a wee bit of the sale – learn a wee bit.' Keeping those precious lists dry and legible must often have been a challenge. A clerk would need an ample supply of initiative, a strong pair of legs and a good oilskin coat with room for himself and his papers. Promotion, when it came, might be a debateable honour as it meant leaving the comparative shelter of the clerk's shed and joining the auctioneer in the sale ring.

'You'd be told you were going to do the "roup roll". That's what they call the book that's kept beside the auctioneer.' The name was centuries old. It referred originally to the sales record of furniture and other items sold at 'roup' or auction when perhaps a householder died or a tradesman went out of business. At the beginning of the 19th century, a number of farmers and estate owners engaged local auctioneers to sell surplus livestock on their own property, rather than negotiating

1. (above) The Highland Drover sculpture outside Dingwall & Highland Marts Ltd. (Terry J. Williams)

2. (left) Drover Ian Munro with his dog, Tan, in 1954. (HLHS. Courtesy of Mr and Mrs I. Munro)

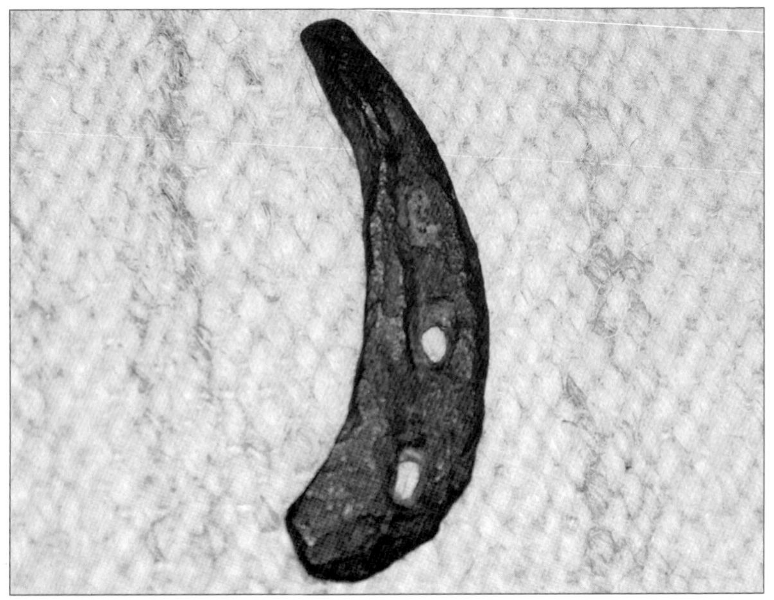

3. The author's talisman cattle shoe. (Terry J. Williams)

4. A campervan perch at the edge of the Atlantic. (Terry J. Williams)

5. Island cattle coming to market on the end of a rope. (HLHS. Courtesy of Robin Valentine)

6. A sale at Milton in South Uist in the 1950s, with George McCallum auctioneering. (HLHS. Courtesy of Robin Valentine)

7. The office car – a scene common to all the island sales in the 1950s and '60s.
(HLHS. Courtesy of Robin Valentine)

8. Buyers from the mainland compare notes in 1957: Ian Oswald, Andrew Hendry and Andrew Binnie with Skye haulier Ewen MacKinnon (far left).
(HLHS. Courtesy of Robin Valentine)

9. Changed days at the site of the Benbecula cattle sales. (Terry J. Williams)

10. All that remains of the pier at Loch Skiport. (Terry J. Williams)

11. Ferryman Ewan Nicolson (second left) prepares to take buyers across the North Ford in 1948. (HLHS. Photographer unknown)

12. Angus Nicolson and Anne Burgess (Ewan's cousin and sister) set out to walk the North Ford in 2013. (Terry J. Williams)

13. Kenneth McKenzie selling cattle at Clachan in North Uist in the 1970s. (HLHS. Photographer unknown)

14. A cattle sale in Dingwall, 2013. (Terry J. Williams)

15. Driving cattle along Dingwall High Street in the 1930s.
(HLHS. Photographer unknown)

16. Modern transport for livestock at Dingwall & Highland Marts Ltd, 2013.
(Terry J. Williams)

personally with dealers at the fairs and trysts. Many of these on-farm sales became annual events and the growing network of railways in the mid 1800s prompted several forward-looking auctioneers to establish auction marts in towns and villages alongside the tracks. By the end of the century, there were over 200 Scottish marts and every auctioneer at every sale would have the roup roll clerk at his side.

'It was a very important job,' said Kenny, 'because the whole price, if you get it wrong, it's wrong all the way through. If the hammer comes down at £100 and you write £200, well somebody would notice. I used to enjoy being roup roll clerk, it was the side of the business I wanted to be in. It was good.'

That was as close as an aspiring auctioneer came to the actual exchange of cash. When Kenny had written down the prices, he would hand over to the 'office'. In 1965, this could have been George Tait.

'The first time I went to Uist, I went as a bookkeeper-cum-cashier-cum-clerk,' said George. 'The office was in the back seat of a car and, to me, that was pretty awful. All day in the back seat of a car, writing. That's how it was. And – I'm sure this is correct – the seller got a slip of paper from us, from the car – so much for the cattle less commission, etc. He took that slip of paper to the car in front, which was the banker, and the banker handed out the cash. In the islands at that time, it was all cash – no cheques. Thousands of pounds – a whole day's turnover.'

George worried about all that cash in the back of a car but Ian Munro pointed out that any thief would have had to get off the island with his loot – no easy task in a place where everyone knew everyone else and the means of escape was an infrequent ferry service with an observant crew. A stranger with bulging pockets would soon find the local policeman taking an interest in him.

While George was confined to the office car, Kenny would be crouching with his roup roll next to the auctioneer, with a view of little more than legs, sticks and cattle tails, keeping track of the bidding, listening hard for the final price, writing it down. Then the next beast and the next until they were all sold; and on to the bus to do it all again at the next stance. Whatever the weather.

'It was all open air in the islands,' said Kenny. 'Stoneybridge had thick walls – it was made with stone. A lot of the other ones were made with timber. We used to take the timber out there.'

This was a valuable contribution. The Atlantic mix of storm-force winds and flying salt spray has always been a powerful inhibitor of timber production in the Western Isles. But the company's intentions were not simply altruistic. Without pens to hold the cattle, their homing instinct would soon have led to a chaos of barking dogs, fleeing beasts and exasperated drovers, who faced trials enough once the sale was over, holding the drove together on the road to the next stance. Despite such precautions, there were times when the day's schedule went awry.

'I can remember once at Stoneybridge and we were late,' said Kenny. 'We didn't start the sale until about six or seven at night and it was getting dark. By the time we were halfway through the sale, it was pitch black. I was the only one with a torch. I was doing the roup roll and Neil Jackson of Corson's was selling. It was a big torch. I would shine the torch on the cattle, then I would take it for me to do the clerking and the buyers were shouting the bids. They always say, if you buy in the moonlight, you've got to sell in the moonlight because cattle always look bigger in the twilight. The dearest cattle we sold that week were the cattle at Stoneybridge. And they probably wouldn't get another twilight sale for selling them. But everyone felt sorry for them – "Oh, better give them a hand, been standing here all day waiting for us to come."'

At some of the more rudimentary stances, the sale ring itself might be little more than a rough space formed by a circle of buyers, sellers and lookers-on, with a basic shelter for the auctioneer.

'Benbecula,' said Kenny; 'I remember you could still see the canopy where the auctioneer stood, just before you come to the crossroads; up on your left, there's a grassy bit which is just on the edge of the big quarry.'

I didn't find the canopy. Maybe it had collapsed into the grass before I got there; maybe the Market Stance Waste Transfer Station had consumed it. Kenny and Ian and Neil and Simon had witnessed the last days of Stansa na Fèille Beinn na Faoghla – Benbecula's Market Stance. Now only the name remained.

For 200 years or more, people had gathered here to trade cattle, horses, sheep, whatever could be bought or sold. Long before the rubbish tip or the quarry or the auctioneers with their protective canopies and roup roll and the clerk in his shed, a drover-dealer might stroll through the crowds, glancing at this beast and that. A crofter's hand might tighten on the rope attached to his one hope of paying the year's rent. If the dealer paused for a longer look, approached to inspect the animal at closer quarters, the crofter would maybe praise his young stirk, pointing out how well it would flourish on the mainland grazings. Gradually, a ring of sorts would form as onlookers gathered to watch the bargaining – the offer, the response, the shaking of heads, the walking away, the turning back, the final slap of hands that signified agreement, a deal done – and the ring would melt away, to re-form elsewhere as the dealer moved from one possible purchase to the next.

While most of the old drover-dealers with their three coats and many pockets would have been trustworthy, they were also shrewd men of business. They would always have the upper hand when it came to bargaining with a crofter who had no option but to sell. The auctioneers brought a measure of stability by creating a single sale ring and ensuring fair play between buyer and seller but none of this happened overnight. Old and new ways of trading probably continued side by side for several years and there would be plenty scope for a canny dealer. With a bit of nimble timing, he could buy cattle on his own terms at one of the local fairs, then sell them on through an auction sale the following day.

Slowly but surely, the small fairs dwindled as the auction companies flourished. By the 1950s, when Ian Munro first came to the islands, the eight twice-yearly sales, at Benbecula, Iochdar, Grogary, Stoneybridge, Carsaval, Milton, Clachan and Ahmore, were well established. The cattle arrived on the end of a rope, the dealers came on a bus to buy them, the auctioneer sold them from under his canopy and the drovers walked them in hundreds to the boats. Having progressed from the clerk's shed to the roup roll, it was in Uist that Kenny McKenzie got his first taste of auctioneering. Ian remembered the occasion well.

'He was a junior auctioneer then,' he said. 'George McCallum was the head of Reith and Andersons and Mr McCallum said to me, "Now we'll give Kenny a shot of selling cattle, so if there's anything knocked down to you don't look amazed." But Kenny started selling and he's been selling ever since, no bother. He learned very quickly.'

'Each company would have had two auctioneers,' said Kenny, 'but [they] did more than sell. If you weren't selling, you were a yardsman or you were a clerk. Everybody chipped in and everybody helped. That's the way it's been all my life.'

Many youngsters went to work at the mart in those days but Kenny was one of the few who found themselves standing on the auctioneer's rostrum, hammer in hand, with a weight of responsibility that would last until the day they retired. The mart attracts fewer trainees these days but, for those who come, the essentials of the job remain the same and the best way to learn, according to Kenny, is still the way he did it – by working day-to-day in every part of the mart. There was no official training scheme, he said, but it took eight years to become fully accredited auctioneer.

'I joined the Institute of Auctioneers with a placement in Scotland and I was three years as a student, five years as an associate and then it was a verbal exam. I went to Aberdeen and there were two auctioneers and they questioned me for twenty minutes. They'd say, "Aye this guy's OK!" or maybe "not OK." And that was you. You were a Fellow of the Institute of Auctioneers.'

The Institute of Auctioneers and Appraisers in Scotland was founded in 1926 as a professional body registered with the Board of Trade and embracing all aspects of auctioneering – from finance to transport, protection of the auction mart system and the training of young auctioneers. Presidents of the Institute were drawn from marts all over the country. The list includes, between 1974 and 1976, George McCallum of Dingwall. FIA Scot is still the coveted accreditation for auctioneers north of the Border but there have been significant changes to the means of achieving it.

'There's university standards nowadays,' said Kenny. 'The young ones have got to go away for two or three weeks a year to do university courses. Then they've got to do a lot of home study. Dingwall & Highland Marts Ltd are attached now to a university down in England and they all go to the same place. This is a very practical, hands-on type of job. We're looking for practical guys who are happy to chase animals, do a bit of computer work or whatever they have to do. Then it just grows and they learn the whole thing from the bottom up. If you send them away to university, the person that comes back with a university degree after five years or so is not really much use to us, to be honest. It's a physical job, you're on your legs a lot of the day. Six o'clock in the morning, the sheep come in off the fields and it's, "You be there." Then when the sale's finished at six o'clock at night, you go and push sheep around, help the yardsmen to get washing – you keep going until the job's done. It's the only way you can make auctioneers of them. Then, if they're interested and keen, they can go round and start visiting. Because canvassing is an important part of the job.'

Canvassing involves calling on farmers and encouraging them to bring their stock to your mart rather than another. In the days when every town had two or even more auction marts, there was great competition for the local custom. Dingwall had Hamilton's mart as well as that of Reith and Anderson.

'There were customers who went to Hamilton's mart and would never come to Reith and Andersons and probably the other way round,' said Ian Munro.

When Duncan Stewart brought Kenny Macrae's cattle to town, it was a Hamilton's auctioneer who sold them. Today, with fewer marts to choose from, a farmer might favour his nearest centre most of the time. But a special sale – a breed sale, for example, or a special sale of bulls or rams – will draw customers from further afield and it could be worth travelling in the hope of a good sale. A mart can become known for certain events, as did Lairg for sheep and Perth for its bull sales (now moved to Stirling).

An Easter Ross farmer might send his cattle to Dingwall and take his sheep to Lairg. Crofters from Skye could take their calves to Portree, their older cattle to Dingwall, and maybe try some sheep at Fort William – all operated by Dingwall & Highland Marts Ltd but each with its own regular group of buyers and sellers. Of the two Uist marts, Ena Macdonald in North Uist favoured Lochboisdale (Corson) over the nearer mart in Lochmaddy (Reith and Anderson) for her pure Highlanders and today will send them straight to Oban (Caledonian Marts of Stirling) for the annual show and sale of the Highland Cattle Society, which draws buyers from all over the world. No amount of canvassing will change Ena's practice. The Skye crofters are Dingwall customers wherever they sell but the sheep farmer of Easter Ross just might be coaxed away from Lairg . . .

This contact with customers is one of the most important aspects of auctioneering. During the two or three years spent preparing this book, I was able to watch the progress of two of Kenny's trainees. As well as working in the pens and helping in the ring and spending time in the office, they mingled with the customers in the cafe, greeting sellers and buyers and their families as well as the local regulars who like to keep in touch with their agricultural roots. Farming is an intimate community – a way of life rather than a job. Sale days offer the chance to meet neighbours from near and far, catch up on news and discuss current events. The mart personnel are part of that community and it's not unusual, in the lull before a sale starts, to see an auctioneer deep in discussion with a couple of farmers at one of the cafe tables.

'You've got to be able to communicate –' said Kenny 'get to know the customers on an individual basis when you're working with them over the years.' I sensed a kind of pride as he talked about the two youngsters and I thought of George McCallum watching the young Kenny McKenzie grow into the job.

As well as his office and financial skills, George Tait was also an accredited auctioneer. 'I'm a farmer's son,' he said, 'and I suppose the interest was there for livestock.' He started work 'with John Swan and Sons in East Lothian, on fifth July 1954 at East Linton auction mart, two and a half miles from home. First you were an office boy. Eventually they trained you to be a typist – sort of. Then you were a clerk behind the auctioneer for a number of years. Then, one day, the pig auctioneer's mother died and there was an auctioneer short, so I was told, "Right, you're selling pigs today." And that was my first. I did practise, walking round the fields shouting out loud, but nobody sits with you and tells you how to say these things. Either it grows on you or it doesn't. Other chaps don't like auctioneering, they'd rather be a cashier, away out of the firing line a wee bitty, you know.'

'It was always quite a thing,' said Kenny. 'If you got a job in the mart, that was your career, like getting a job in a bank – the second son on the farm used to go and work in the mart and could be a clerk or an auctioneer.' And for those who found a better fit down another road, their time with the auctioneers stood them in good stead.

'Today for example,' said Kenny, when we met in his office at the Dingwall mart, 'of the three or four sheep buyers we have in the ring today, two of them started their career working for Aberdeen mart and they've gone on and got a job. One works for Morrisons the supermarket but he did his basic training with the auctioneers. The other chap was the same – he was some years with the mart in Aberdeen.'

Aberdeen was, and is, one of Britain's largest livestock centres. In the late 19th century, the city was home to no less than four auctioneering companies (including Reith and Anderson) and four auction marts – Central, City, Belmont and Kittybrewster. In addition, the Agricultural Hall hosted sales of pedigree cattle and horses and down at the harbour was a sixth auction where livestock from Orkney and Shetland were sold. Over the years, some of these marts were taken over or closed as one company merged with or sold out to another. In 1947, a final amalgamation led to the formation of Aberdeen and Northern Marts, with no less than 13 sale rings strewn across the busy city.

'I can remember when we did the auctioneers' conference,' said Kenny. 'Aberdeen had three busloads of clerical staff and young auctioneers – a huge firm. Nowadays, they don't have anything like that. They're all centred in Thainstone.' This massive auction centre opened in 1990 on a 500-acre site near Inverurie, 16 miles north-west of Aberdeen. It brought all the city's marts together in one impressive complex, which is today considered the most modern auction mart in Europe. Before that, said Kenny, 'there might have been seventy or eighty youngsters in the Aberdeen markets but only maybe two or three made it as actual auctioneers for a whole career. It's difficult to get youngsters to go into it nowadays. There's not nearly so many of them – not so many opportunities. The whole education system has made it more difficult for the auctioneering firms to get youngsters. Nowadays, unless you go to university, you don't do anything. That is where I think it's all lost as far as we're concerned.'

Whatever else might change, the auctioneer's task has remained unaltered.

'You're estimating the value,' said George. 'You're looking for the right buyer. You've to try and hit the middle. It doesn't always work but you've a good idea – a lot of buyers always buy a similar type of animal so you know, if it's a poor one, that half-dozen people might bid on it. If it's a tremendous one, that different half-dozen people will bid. You need to like people, yes, and they need to like you. None of us are perfect but it just grows. If you're quite keen, ach, you make the grade.'

'Encouraging people to buy,' said Kenny. 'You've got to keep the sale moving. You've got to try and humour customers. They might not be happy. Some will stand at the gates so the beast won't leave the ring when the hammer's down. Things like that annoy you but you just have to go with it. You must keep the sale flowing.'

'There's always work the day before,' said George. 'You've got to prepare for the sale – you know, cataloguing and getting a sort of order. And, in lots of cases, phoning up buyers and sometimes organising transport. So an auctioneer's not just a guy who goes into the rostrum and shouts. It's a good life. I was with three different firms in my career – Edinburgh, Dingwall and Inverness. I've no regrets.'

After his Western Isles initiation in the office car, George was assigned to the auctioneer's rostrum at the company's sales in the Isle of Skye. Kenny helped keep the Uist sales flowing for over 40 years and he looked back fondly on the days of travelling from one market stance to the next through the Uists and Benbecula.

'It was a lot of hard work but there was a party every night,' he said. 'I left on the Friday morning and I came home the following week, Saturday. So it was eight days away on the island. Ach, it was good fun.'

# 12

# *Not Only Cattle*

Browsing through my notebook, I found the very first telephone number I'd written down at the Drovers' Tryst in Crieff. Essie Stewart had invited me to visit her at home in Sutherland and the sheep sales at Lairg were imminent. The journey from Dingwall wouldn't take more than an hour or so. It was time for another diversion. Essie answered the phone. Oh, yes, she'd be delighted to see me; she'd put the kettle on.

The Stewarts of Lairg were one of the oldest travelling families in Sutherland. Essie was born in 1941 and she spent almost 50 summers on the road. During her childhood winters she lived with her mother and grandfather in one of three small houses built on high ground a few miles outside the town. There were Stewarts in both the other houses too.

Every spring, the horses would be harnessed, the carts loaded, the houses closed up and, for the next five or six months, home would be a 'bow tent' – a frame of hazelwood rods covered with tarpaulin. It could be a tough life, exposed to all weathers and plenty hard work but there was a freedom in it – the fresh air and the mountains and the warmth of friendship. Often, several families would meet up at the camping places and share music, songs and stories, all passed down by word of mouth from one generation to another. Essie's grandfather, blind Alexander Stewart – known as *Ali Dall* in his native Gaelic – was reputed to be one of the greatest of all traveller storytellers. He had learned most of his stories from his mother and he gave them, in turn, to his granddaughter.

From Lairg, roads fanned out into the remote and rugged landscapes of the far north-west. There was the westerly route by Oykel Bridge to Elphin, Ullapool, Lochinver; or the carts might head north-west along Loch Shin into the mountain country of Arkle and Foinaven, to reach the coast at Loch Laxford. But it was along the road that led almost due north, through Crask and Altnaharra to Tongue on the Pentland Firth, that Essie remembered meeting the drovers.

'Not cattle – sheep,' she said, 'coming to the sales at Lairg in August. The Tongue road, when you go to the Crask and you start going down Strath Vagastie, it's a long stretch. You look from the top of the hill and you can see right down to Altnaharra and Strathnaver and I remember that glen just being both sides covered in a mass of white sheep. And their lambs as well. It was a wonderful sight, wonderful. Because there was two or three sheep stock clubs. You had the Altnaharra people and you had the Strathnaver people and you had the Tongue people. The shepherds had to keep the sheep separate and you would get this sort of gap in between. Thousands – and we are talking thousands – of sheep and lambs. And then you would get the gap. The dogs barking, shepherds whistling and shouting to each other, coming down. And we just pulled off the road and it took a long, long time for them to pass. These shepherds that we knew and the banter that went on between them . . . It was a fantastic sight. Every time I travel that road, when I come round the corner and look down, in my mind's eye, I can still see that sight. We are talking sheep for miles, miles, on either side of that valley.'

The Lairg lamb sales have laid claim to be the biggest in Europe. At one time, as many as 40,000 sheep would go through the sale ring in one day. There used to be cattle sales and horses too. The travelling people were great horse traders. Horses were in their blood. The Stewarts would buy horses in the west to sell at the big autumn sale in Lairg and Essie would help to ride them home. She loved the horses and couldn't resist the sales.

'Jouked the day off school, didn't tell my mother and she always found out. There would have been a group of us. It was all planned the day before – "Are you going to the sale?" "Are you?" "Yes, well, right that's fine." We would meet at the mart. It was such fun and then my mother would hear about it – "You were at the sale? You were told not to go." Obviously she knew. She could suss it out – anxious to get away, the morning of the sale. Other mornings, dragging heels but the morning of the sale – up, chirpy, whistling.

'I loved being there when the horses . . . I remember we used to sell horses as well and after the ponies would be sold – or pony, depending on how many – used to walk them down to the railway station and box them. And my grandfather always made sure that the horse was boxed properly. Yes, a lot of the cattle went on the train. There were special trains put on for the cattle, for the sale. We lived quite close to the railway and extra goods trains would be put on and my grandfather would say, "Ah, that's the cattle trains." So, yes, it was by road and by rail.

'In those days the Lairg cattle sales were huge, huge. And

being in that sale ring and listening to that auctioneer and trying to mimic him. And not only were you going and having fun, you were meeting people that you knew. People coming to chat – "Does your mother know that you're here?" "No? Well, you'd better come and have a cup of tea!" There was a long wooden shed at the side of the road and the cafe was there and thick white mugs – mugs that would hold probably half a gallon. But the tea tasted so good. Maybe because the mug was white, I don't know. Isn't that funny how things like that stick in your mind?'

While Essie was enjoying her tea, young Alasdair Cameron from Ross-shire – treated to a day at the sales with his father – was counting beer-bottle caps. Thousands of them, removed and dropped under a makeshift bar year upon year, in the attempt to quench the great thirst of the Lairg sales. Alasdair noticed that the newest, shiniest layer rested on other deeper, rustier strata in a kind of socio-agricultural geology.

I came late to the mart at Lairg. Too late for the horses and cattle, the thick white mugs of tea and the layers of bottle caps (long since cleared away). But not too late for the astonishing sight of 17,000 sheep in one place at one time. Not too late to lean on the elbow-polished rail round the sale ring, in the wooden building with its curved tin roof. Not too late to watch, to listen.

Sheep everywhere – bleating, bawling sheep. Waterfalls of sheep pouring out of triple-deck lorries from Inverness, Aberdeen, Cumbria, Kelso, Caithness. Streams of sheep flowing along intricate cross-hatched alleys, grass turning to mud under thousands of cloven feet and hundreds of no-nonsense boots. Sheep marked with dots, circles or letters in red, blue, green or black, on rump, shoulder, nape or flank.

Sheep clustered in 800 lichen-bearded pens embossing the hillside.

'Hup, hup!' Men keeping sheep on the move. Men counting sheep – out of the lorries and into the pens; out of the pens and into the sale ring; out of the sale ring and back to the pens; out of the pens and into the lorries. And, much later, at the end of a long road and a long day, out of the lorries and on to the farms. Counting sheep – it's a wonder anyone is still awake.

Waiting for the sale to start, sellers check the catalogue and cast an eye over the competition. The talk is of the weather, the shearing, the hay and sheep.

'There's a good bloom on the lambs this year.'

'They're as good as they'll be. It's up to someone else now to make them better.'

Among the pens with my camera, I fall into conversation. Mr Mackay – tweeds, shepherd's crook in hand – tells me his father and grandfather sold sheep at Lairg before him. He's getting on now himself – not as able as he was. Does he have anyone to follow him into the farm? A doubtful look. But, yes, he's got lambs here to sell today.

The crackle of a tannoy echoes across the pens. It's started. Inside the sale ring, the white-coated auctioneer faces a full house from his pulpit-like platform. The day's aspirations of every seller depend on his performance.

'Come in, Mr MacDonald, how's yourself today?'

It seems Mr MacDonald would like £60 apiece for his lambs. The auctioneer tries his best to oblige.

'Look at these lambs – they're absolute belters. Come on now, great lambs, what will you give me?'

The bids come with a nod, a raised eyebrow, the twitch

of a finger. The auctioneer spots them all, pits one against the other, eases up the price. As bidding slows, he persists, reluctant to bring down the gavel.

'Any more? Last chance . . . all away now . . . going . . . going . . .' Crack! Mr MacDonald's lambs are sold for £57 each. Not bad. Next please.

Selling continues without a break. Outside, the sheep are still arriving. A sudden thunderstorm – flashes of lighting and rain pounding on the corrugated iron roof – merits no more than a brief aside to the crowd.

'Don't worry, the ring is a wooden building.' Next please.

Two truck drivers, chatting outside the ring, blue boiler suits embroidered with their company name, have brought sheep from 'every nook and cranny' of the country along their route. The new drovers.

'There's not the sheep that used to be in it,' says one. He's seen forty thousand sold here in one day. 'You just kept going until the job was done,' he says. Sheep collected, sheep sold, sheep delivered, finish at three o'clock the next morning. You can't do that now. You're restricted to ten hours driving (plus breaks) and fifteen in total per day, which means two drivers are needed for long runs. And you can only do a one-way trip at Lairg – deliver or collect but not both. The lorries have to be washed out between loads and this old mart doesn't have the facilities.

A small boy passes, purposeful in work boots, waterproof leggings and the smart green sweatshirt of Wm Armstrong, Longtown, Cumbria – a ten-year-old apprentice drover. Armstrong was one of the major cattle-reiving names of the Borders, way back in the 15th and 16th centuries. Now it belongs to one of the biggest livestock hauliers in Scotland.

The truck driver is gloomy about the future. He's seen a downturn in cattle and sheep over the years. Every hill used to be full of sheep. It's trees now. An oblivious young Armstrong makes his confident way through the maze of pens.

As the balance tips in favour of sold over unsold, R. W. Stewart of Bo'ness pulls into the loading bay. Blue triple-deck lorry, two drivers with sweatshirts to match the gleaming paintwork, immaculate sawdusted accommodation for the cloven-hooved. Their consignment of lambs, to be delivered to new homes for fattening, is urged down from the pens by men in blue coats. United Auctions, who run this and several other Scottish marts, have sent up a squad for the day from Stirling and Aberdeen. Could they be called drovers too? A chuckle. Actually, one of their team did work as a drover out in the Uists at one time. He's not here today but I'd find him at the Dalmally mart in Argyll, if I would take a run over that way.

# 13

# *Dealing and Wheeling*

'I've never in my life experienced midges like it,' said Kenny. 'I can remember the waves of midges. You were choking on the midges, there were that many of them.'

I lifted a photograph out of the archive. It showed part of an island sale ring, the crowd standing three or four deep, with a line of raincoated and tweed-capped dealers at the front, every one of them rubbing an ear or scratching his head. Kenny knew better than to take their gestures as offers to buy. He was suffering along with the rest. Bids, when they were deliberately made, could be subtle – a raised finger, a nod of the head, even just a glance. The auctioneer knew the regular buyers well, both local and mainland, and he would be looking out for their particular ways of catching his attention.

Simon and Neil Campbell remembered without hesitation the names of dealers who had come to the islands all of fifty years ago: Andrew Binnie, Andrew Hendry from Stirling; Willie Hendry, Ian Oswald, Tom Adam, Bob Love, Bob Shaw from Inverness; Donald Maclean near Muir of Ord. All these came out on the boats from the mainland but there were local buyers too, said Simon.

'There was no calves going away yon time, like today they're going at six, eight months. They were keeping them. One farmer or one crofter was buying the yearlings and he would sell them to the people of the mainland at two years and three years. They were no use to the mainland buyers because they had to keep them for a couple of years before they would sell them to the butchers.' He pointed to a photograph of a sale at Milton.

'That's John Macmillan's grandfather,' he said. 'He was a John as well. He was always standing behind the auctioneer when he was buying cattle. With his stick, he would go like this [Simon tapped me on the shin] and nobody else was seeing until they would hear, "Sold to John Macmillan, Milton." That's why he was standing there with his stick!'

Another photograph, another local bidder. 'He was from Iochdar – he used to buy quite a few.'

Neil told a story of one of the mainland dealers checking his cattle before they were loaded on to the boat at Lochboisdale.

'I remember the day I was gathering with young Ronald MacCormick, Lochboisdale,' he said. 'We were gathering Donald Maclean's cattle. And he came off the boat at Lochboisdale. He walked right through the cattle there. He says to Ronald, "Where did you get that beast?" and he pointed to a beast there. "Well I didn't buy that beast, they must have swapped it." Out of three hundred beasts. Oh, they had an eye for buying cattle, yes.'

From the minute the hammer came down in the sale ring, each animal was linked by its paint mark and scissor clip to the dealer who had bought it and woe betide any paint person or scissor clipper who got the mark wrong.

'Everyone was very touchy about what mark they had,' said Kenny. 'You just had to get that mark, then the clip mark, maybe four or five inches long, just in case the paint came off. By the end of the week there, you'd have sold maybe two thousand cattle. They were all at various times getting mixed together so . . .'

The drovers got Donald's cattle on board, 'travelling in the old *Hebrides*,' said Neil. 'They sailed, they left Lochboisdale – quite choppy at the time. It got worse and the captain says to Donald, "Will I turn back to Lochboisdale?" "No," he says, "if the boat goes down, all my money's aboard here and I'm going down with it."'

Generation after generation down the centuries, that was the risk the livestock dealers took. They put what money they had into the best stock they could afford, in the hope of selling at a profit so they could buy more and better next time. Many must have fallen by the wayside but some of those who succeeded took on legendary status.

There was John Cameron of Corriechoille in Lochaber, held to be the greatest drover in Scotland in the 19th century. He began his career as a young lad in stocking feet, driving cattle and sheep to market for some of the main dealers of the time. By 1840, through careful saving and trading, the poverty-stricken crofter's son was mounted on horseback and reputed to be sending thousands of cattle south to the Falkirk Tryst in droves that stretched over seven miles.

A century later, Kenny Macrae from Ullapool – Kenny The Drover, who employed Duncan Stewart to drive his cattle to the Dingwall mart – was recognised as the most successful drover and cattle dealer in the north of Scotland. Duncan's drove over the Dirie Mor was a tiny part of Kenny's enterprise. The dealer would walk the length of the islands, buying cattle and bringing them in their hundreds across the Minch to

Ullapool. Everybody knew Kenny; he was liked and respected wherever he went.

The next generation of Uist dealers travelled through the islands by bus. The mode of transport may have changed but the road to success was no different from that of their predecessors.

'When Bob Love [first] came here,' said Neil, 'he was nineteen or something. He only managed to buy three cattle; that's all he bought. The money wasn't there. Oh nice man, he was coming here for years and years and years.'

Persistence and canny dealing paid off but character counted for a lot. The dealers who came back every spring and autumn were a welcome and much-appreciated presence, not just for economic reasons.

'They would talk to you, even when I was young,' said Neil. 'Andrew Hendry was telling me he was never in bed after four o'clock in the morning. Seven days a week, three hundred and sixty-five days a year. Every day, he says, feeding cattle. And he wasn't selling in the Stirling mart at all. He was getting phone calls from England: "I want ten or twenty or whatever; just price them." They didn't even ask, "How much are you going to charge?"'

Trust. Without it, the droving trade could never have survived.

'Come and I'll show you a photograph,' said Ian Munro one day. We had been watching one of Dingwall's regular buyers bidding for sheep. Ian led me through the Droving Exhibition. 'That's Bob Shaw,' he said. Mr Shaw had been a very popular dealer in the islands and it was his son who had caught my attention at the ringside. I noticed that he always stood in the same place.

'Yes,' said Ian. 'Look round the ring and you'll see that all the buyers have their favourite spot. It's so they can compare each animal from the same point of view at every sale.' The hammer fell and the auctioneer announced, 'Shaw three.' I was puzzled. Was this some secret code used by the dealers?

'Dealers?' said Kenny McKenzie. 'Dealers is something of the past. A dealer used to buy stock and he would try to sell them at another market. I remember right up until the 1990s, half a dozen guys came from Aberdeenshire. They were dealers. They bought, took them in a lorry back to Aberdeen and hoped to sell them on the Friday for a profit. But that doesn't happen any more. It's livestock agents nowadays. An agent will have accounts for several different people. Then he sorts it all out in the office and he tells us where they're going. He may still pay but he's not dealing. There's a couple of guys that buy for themselves but they can't put them straight to market. They'll maybe take them home, have them for the relevant time, then they'll put them into another market. The thirty-day restrictions. Foot and mouth put paid to all the dealer stuff.

'A lot of farmers are one-man-bands nowadays. They haven't got time to come to the mart so they're quite happy to get Mr Shaw, for example. He'll go and he'll buy for number one, two, three, four . . .' Each number referred to a specific farm. The paint marks and scissor clips had been replaced by numbers and I was right – it was a sort of code.

'Way back, it was just all different people buying for themselves,' said Kenny. 'You could get about fifteen or twenty people going out to the islands to buy. Now you're lucky if you get three or four to go out. Farming now – there's less people with stock. You travel around and maybe only one in ten farms has stock. The farmers themselves have to work. Before, they maybe brought families up on a sixty-acre holding. Now, you'd need six hundred or eight hundred acres to bring a family up. It's not easy.

'We had a sale in Dingwall called the West Highland Sale and I think it was six or eight people used to sell at that sale. They were all dealers but they used to go round all these sales in Uist and they'd go to Lewis and Skye and they would take grazings and keep the cattle. So they used to buy a lot of cattle in summer and then, for the one particular sale in the back end, there might have been fifteen hundred cattle, with just six or eight people. Some of them would have maybe two hundred cattle.'

Back at the ringside, the auctioneer had set a brisk pace and the sale was going well with a steady stream of bids for each lot. I began to work out who was a farmer buying for himself and who were the livestock agents. It was fun to try and guess each client's preferences from the type of animal the agent was buying. No only was he responsible for selecting the right stock but he had also to decide exactly how much of his client's money he was prepared to spend. Bidding hard and fast – a slight nod, a lift of the head, a hand or a finger – he stopped abruptly as soon as the price went above his limit. There it was again, that vital element of the droving trade – trust.

The hammer came down once more and the auctioneer called out, 'Shaw six!' How did he know? There was too much noise for verbal communication across the ring. Was lip reading part of an auctioneer's training? I watched carefully as the next lot went to 'Shaw eight!' I saw the buyer place one fist briefly above the other – a drover's version of sign language!

The exit gate opened and the bewildered sheep hurried out.

The cry of 'Shaw eight!' reached the head yardsman standing at the entrance to the pens behind the sale ring, all under cover and echoing to the din of clashing metal gates and bawling, bleating livestock. The yardsman checked a printed list, raised one arm and signalled a number to the first of several men waiting to guide the sheep to one specific pen out of hundreds. He, in turn, signalled to the next man and sent the sheep scurrying on their way. Left, right, straight on and left again, they were passed from one gesticulating worker to the next until they reached the pen allocated to Mr Shaw's client number eight, the gate clanged shut behind them, and I was left wondering how long it would take to master the drovers' sign language.

What about the roup roll? It was still there, next to the auctioneer – a computer with the roup roll clerk typing in the details of every sale, printing out sales sheets and handing them to a colleague who took them into the mart office where, after more computer calculations, the buyer would pay for his purchase and the seller receive his payment. It was a far cry from Kenny in his oilskins and George Tait in the back seat of the office car. Much had changed since Bob Shaw's cattle were walked through the Uists by Ian Munro and his fellow drovers, to the sound of weavers' looms and the thunder of the Atlantic. The sheep bought here today would arrive with their new owners after a journey of a few hours in a fast, modern, articulated livestock lorry.

'Cattle movement follows the same routes,' said Kenny McKenzie. 'But it's all done by lorry now.' His working life had spanned a remarkable period in the history of livestock in the Uists and Benbecula.

'The first year I went over, 1965, that was the last year that we did the complete droving,' he said. 'Nineteen sixty-six would be the first year that George McCallum took the lorries across and we did some droving and some shifting with the floats [livestock lorries].'

The Campbell brothers also witnessed the beginning of the end of walking with the cattle. Were they still helping when the floats came?

'Yes, yes,' said Simon, 'loading from the sales, from the wee sales. No walking then, no. And the floats just going straight on to the boat. There was a side ramp on the boats yon time. You would put them on the side ramp. The cars would go in on the side ramp too.' These were the hydraulic ramps of David MacBrayne's new vehicle ferries, which transformed transport to and from the Western Isles. And did the new methods save time? 'Och, you would be at them all day, even with the floats,' said Simon.

Cattle destined for Oban might still have been walking to the boats at Lochboisdale at that time, Kenny thought, but, when other hauliers started to come out to the islands, Reith and Anderson decided to take all their cattle from the sales to the Lochmaddy pier head. Ewen Mackinnon from Skye, known to everyone as Ewen Crossal after the place where he lived and ran his haulage business, was one of the first to follow George McCallum's lead.

'After 1965–66, Ewen Crossal had lorries,' said Kenny. 'They weren't big lorries; they would be twenty-foot containers and they might have taken fifteen or twenty in each one. The lorries would cross from Uig in Skye to Lochmaddy and go down to the sales in South Uist.

'We would shift everything up to Lochmaddy,' said Kenny. 'It would be a three-hour round trip, to go from South Uist up to Lochmaddy and back again on single track roads. They would load twenty cattle and go away up to Lochmaddy. Then they would come back down to another sale and take another load and it was a long day for these guys. But we had good penning facilities in Lochmaddy.

'So then we would have to go up and sort them into various lots. I was starting to organise things that we could try and get the stock to go direct on to the lorries – fifteen there for Bob Shaw, or whatever, so they weren't mixed the same as in the days when they would take the drove to the pier head and then they were all loaded [together].'

Gradually, one or two enterprising islanders bought their own lorries and local hauliers became available. One was Alasdair Macdonald of Ahmore in North Uist, where the last sale of the drovers' week was held. Today, Alasdair's lorries bear the sign – 'A. Macdonald & Son Ltd.' – and they carry many more than the 15 or 20 beasts of Ewen Crossal's original floats.

On the mainland, some drover-dealers had already invested in motorised transport. Kenny Macrae bought a three-ton Bedford lorry in 1939 and, a few years later, he acquired a two-deck, five-ton Austin. There were no ferries to take his lorries out to the islands at that time and he continued to move stock on foot but he certainly anticipated the changes that were to come.

A long way south, just across the Scottish–English border in Longtown, Cumbria, William Armstrong bought a Model T Ford in 1927. It could carry just one well-behaved cow but it was a start. In 1936, he put his first fleet on the road – two Bedford trucks, one new and one second-hand. By the time I found myself watching a small boy in his smart green Armstrong sweatshirt stride confidently through the mart at

Lairg, Wm Armstrong (Longtown) Ltd had, for many years, been one of the largest livestock haulage companies in the country.

I wondered if that small drover boy knew he was one of a long line of livestock shifters? Had he heard of a lad who, in 1542, at 12 years old, was believed to have fought with the English at the battle of Solway Moss? Kinmont Willie Armstrong, as the young fighter became known in later life, was probably the most infamous member of the most notorious family of Border reivers. Cattle, sheep, goats, horses – all were lifted – and from either side of the Border. Family came first in Border warfare during the 15th and 16th centuries. An Armstrong would aim to be on the winning side, not caring whether it were Scots or English. They were not the only reivers – there were Elliots and Kers, Maxwells, Johnstones, Hendersons and Irvines. When a kind of peace and a legitimate droving trade emerged in the 17th century, there were surely Armstrongs among the drovers and dealers. Skills inherited from their cattle-rustling predecessors would serve them well.

Wm Armstrong Ltd remains a family business today and, from the Pentland Firth to far south beyond the Border country, Armstrongs are still shifting cattle. Of course they are not the only livestock hauliers – there are MacTaggart and Stewart and their island counterparts, Macdonald and MacAskill. These are some of the 'big operators' with their 40-foot, double- or triple-decked lorries. There are other smaller, local hauliers but there's no guarantee how long they'll be able to hang on as livestock numbers dwindle.

'We are getting more island stock coming into the Dingwall sales direct,' said Kenny McKenzie. 'Maybe in another ten or twenty years' time, there could be even less people with stock. So the few who do have stock have to fill a lorry and take them down here. That happens all the time from the west coast – Gairloch and Shieldaig and all these places. There used to be sales round these places as well but they've all disappeared – every one.'

In time, the eight sales in the Uists and Benbecula were reduced to two with the construction of purpose-built marts at the pier heads in Lochmaddy and Lochboisdale. After a period of shared responsibility between Reith and Anderson and Thomas Corson, each mart was placed under a single auctioneering company. Since 2002, Dingwall & Highland Marts Ltd has handled the sales in Lochmaddy, while the Lochboisdale mart is run by United Auctions of Stirling. With dwindling attendances and increasing numbers of livestock bypassing the island marts, is it still necessary to go out there? Kenny believes it is.

'We just have to try,' he said. 'The islanders are entitled to a living just the same as everybody else. The way this market is run, we're there to provide a service.'

There may come a time when the Uist crofters say, 'Ach no, we can't do this any longer, we'll just fill a lorry and come down.' That day, if it comes, will mark the end the Uist sales.

# 14

# *Return to Lochmaddy*

It was late April and the Dingwall team were preparing to go over to Lochmaddy for their spring sale of cattle and sheep. I was going to join them but I wanted to spend one more night in the islands and I wanted to be in Lochmaddy when the cattle came in, so I was going out a day early. Once again I packed the van – sleeping bag, camera, notebook . . . and cattle shoe. It had been with me for the whole adventure so far and it was going to share this last stage of the journey.

Growing excitement as the westward road reached the edge of the mainland, soared over the bridge to Skye. Impatience as the ferry ploughed steadily across the Minch. Leaning on the rail, watching the Uist coastline rise, much too slowly, from the horizon. Haste to escape the confines of the vehicle deck, to follow the familiar grey thread past the lochans and peat bogs to the machair and the dunes. A sliver of dry land beside probing fingers of salt water. Sleep and the Atlantic booming. A fine morning and a small breeze, breakfast outside and watching the light change over the sea. It would be easy to linger but I don't want to be late.

At nine o'clock there's an absence of activity at the mart beside the pier. There's time to sort out my sleeping quarters and book a place on the early-morning sailing tomorrow. I want to travel with the drovers and the cattle. Yes, there's room for a small campervan, says the lady in the Calmac office; yes, I can stay overnight in the car park above the pier; and, yes, I can fill up with water from the tap outside the office. Perfect. All we need now is some cattle and a sales team.

A. Macdonald & Son Ltd, Ahmore arrives with the first consignment. Three cows emerge from an enormous lorry. Everything falls quiet again. Three cows and a notice pinned to the pens are the only indication that there's a sale here today. The Dingwall squad will be sailing from Uig in a few minutes; the crossing takes an hour and three quarters. I go to make coffee in the van.

The local yardsman arrives, as does the first 4x4 vehicle towing a stock trailer. The driver stops, checks the notice to see which pen he's been allocated, then manoeuvres to unload a couple of cattle. More follow. Gradually it gets busier. Ahmore comes back, with a full load this time. As he unloads (shouting, clatter, instructions, bellowing), trailers queue for their turn. The doors of the mart building open; there is movement in the office window and conversation around the pens. Gaelic conversation.

The bawling of cattle invades Lochmaddy's everyday routine. Cars arrive and join the ferry queue. The *Hebrides* comes into sight at the far entrance to the loch, creeps slowly towards the pier. Warning bells sound and the ramp clatters down on to the slipway. Foot passengers descend the gangway, the Dingwall team among them – the auctioneer, the clerk, the cashier, the drovers. Livestock trailers are caught up in a flurry of vehicles leaving the ferry and, for a moment, there's a risk of gridlock in Lochmaddy. One long, articulated cattle float –

Longthorne Bros – comes off the boat and pulls up alongside the mart. The driver hands down overnight bags. Contracted by Dingwall & Highland Marts Ltd to help transport cattle from the sale, he's brought the luggage and maybe a couple of the drovers too.

Kenny McKenzie has brought his car across on the boat. He draws up beside the mart building and the office staff are soon busy unpacking computers, record books, everything that ensures the smooth running of a modern livestock sale. Outside, the rest of the team move along the pens, discussing arrangements with the local yardsman. Kenny's most recently qualified auctioneer is doing the selling today. He's looking over the cattle and chatting to the crofters, who have switched seamlessly into English for the benefit of the east-coast mainlanders. There are a few pens of sheep. There are buyers too – some local, some from the mainland, mostly Aberdeenshire I gather. They are inspecting the cattle and sheep, taking note of those that might be worth a bid.

Meanwhile, the ferry is taking on vehicles for the return journey to Skye, and Lochmaddy is full of parked trucks and livestock trailers. Every designated and non-designated parking space is full; so are the pavements and the verges. Crofters lean on the pens, sizing up each other's livestock and enjoying the craic. The sale is a change from the daily round – a chance to meet up and compare notes. Everyone is anxious about prices, hoping they'll be good. There's no shouting now, no rush. Even the animals soon settle. The last few sheep arrive. The ferry hums and sends up gentle puffs of dark smoke.

Precisely at noon, Kenny McKenzie emerges from the office, leans on the rails.

'Right then, folks, if you want to come in, we'll get the sale started.'

The ferry has slipped away unnoticed. Everyone moves into the mart building – a cold, functional place of stone and concrete, metal rails round a small ring and a raised auctioneer's rostrum in the corner with just enough room for the clerk to join him. Behind the rostrum is a hatch through to the office; inside the ring, a wooden step next to the auctioneer where a seller can stand within earshot. At the ringside are a few moveable wooden benches and against the walls two or three tiers of broad wooden steps. Some people are sitting on bags of what looks like cement. Opposite the entrance door is another hatch with a tiny kitchen behind and a member of staff from the Lochmaddy Hotel, who has brought a big pot of soup and the means of providing hot drinks and snacks.

The pace is unhurried. A portable digital screen is brought in and hung on the wall. There's a weighing machine at the entrance to the ring. As the first beast arrives from the pens, Kenny (a drover for the day) calls out its weight to the auctioneer, who taps the figures into a small machine, which displays them on the screen. The youngest member of the team (Dingwall's second auctioneering trainee) shouts out tag numbers, sometimes after a couple of turns round the ring following a beast determined to avoid any examination of its ears. The number is matched up with the animal's passport, which includes details of its birth date.

The auctioneer calls for an opening bid and the atmosphere instantly changes. There's a degree of showmanship involved – the white coat, the rapid-fire tongue-twisting speech, the gesticulating, the suspense as the hammer is raised, all laced with courtesy and humour. The faces round the ring are pictures of concentration, watching everything, giving nothing away. The auctioneer gives them his full professional attention, at top speed, coaxing and cajoling the buyers, praising the cattle, taking instruction from the seller.

'Is that enough?' he asks. The crofter, already grim-faced, scowls even more and the persuasion continues, squeezing bids from buyers who perhaps feel they've already offered more than the beast is worth. The hammer falls at last and the crofter has got a good price, though he's not going to admit it. That's what it's about, no matter where or how big the sale. Selling, selling, until the very last bid is extracted and the hammer cracks down. Just as Kenny said – keeping the sale flowing.

My ears are ringing. The temporary public address system is too powerful for this small building full of echoes. If it's sore on human ears, what must it be like for the cattle? These big continental animals are already wilder than the old Shorthorn-cross-Highlanders, whatever problems those much smaller animals caused their drovers. 'Watch yourselves!' is the cry as a couple of lively stirks thunder out of the ring on their way back to the pens.

The cattle sale has lasted one and a half hours. It's time for the sheep and there's a pause while the tag reader is set in place. Reading half a dozen or more tag numbers on a constantly moving group of small ears hidden by wool is more than any drover can manage, so the sheep pass through a portable crate and their numbers are electronically read and recorded on a laptop computer perched on a wall.

Everyone is cold. Soup and tea are collected and consumed. There's time for a stretching of legs and a blether. The air is full of Gaelic again.

I remember Kenny saying that sheep were not of much interest in the islands in the 1960s. There are not many here today. Yet one man refuses to sell his sheep below £50 apiece. He'll take them home, he says, and I wonder what he'll do with them now. Another seller is less anxious and laughs as her one tiny sheep, all on its own in the ring, is sold for £5. Others reach £60. All are small compared with the mainland flocks; small and sweet perhaps, and ready to grow on when they reach good feeding on the mainland. After half an hour's trading, all the sheep have been through the ring.

'And that concludes today's sale, thank you,' says the auctioneer, taking off his microphone.

Now it's the turn of the office staff to face the public. The door is crammed with people as sellers collect their money and buyers make sure they've been assigned the right beasts. Once that's done, everything will have to be dismantled and packed away again.

The rest of the team are already out among the pens. Stock have been allocated to pens according to who bought them, then into groups with regard to their final destination. Some will go to Dingwall and on to their new owners in another lorry from there. Some may head straight to Aberdeenshire, being dropped off farm by farm. These have to be organised carefully, to avoid unloading the whole consignment in order to deliver a few beasts inconveniently placed at the front.

I lean on the rails, watching the men at work and admiring a fine bullock in a pen of its own. A young crofter approaches. The animal comes up to him and he strokes its forehead.

'Off to a new home?' I ask.

'Off to the slaughterhouse,' replies the crofter. 'That's what it's about.' He pauses, then, 'You get attached to them, you know.'

His bullock made the top price of the sale. The local butcher has paid over £1,000 but I can tell the young man would rather the beast had gone to a mainland buyer, even if the outcome would eventually be the same. At least it would have left the island alive. But this is business and that's how it is.

He walks away and I remember the people who told me about the cattle going to market on the end of a rope, the children crying at saying goodbye to their calf. I remember the drovers' accounts of these 'pets' running away from the drove as they passed their home road end, trying to get back to the place and the people they had known since birth. It was business but a very personal business and my conversation with the young crofter of North Uist has confirmed that here in the islands that personal element survives yet.

The cattle and sheep are settled with a good feed of hay, ready for loading in the morning. The drovers, along with the Longthorne Bros lorry, have gone down to Lochboisdale. Kenny has told me that he arranges transport for both sales now, taking all the livestock off the island via Lochmaddy. The village resumes its peace. A few local buyers come with trailers to collect their purchases. I notice that the man who refused to sell his sheep was quick to load them and leave as soon as the sale ended. I wonder, 'Has he gone south to try for a better price?'

At half past six, there's a burst of shouting, cattle bawling and sheep bleating. Longthorne Bros is back and unloading livestock from the Lochboisdale sale. There's a surprise in this load – two full pens of beautiful little Highlanders. Surely these must be Ena Macdonald's youngsters. I remember her saying that she sends them to Lochboisdale rather than Lochmaddy. It's not a matter of distance (she lives in North Uist) but of reaching the right buyers. These little hairy beasts are going to Stirling, Kenny tells me. And the Stirling buyers will have come over with the United Auctions team. It all fits together, like pieces of a jigsaw. Here are crofters doing what they have always done – looking for the best price; going from one sale to another to try for a better deal; taking their animals further to connect with the buyers they know and who know their stock. It's not just tradition for the sake of it. It's the way they've done it for a long, long time because it works.

The day's business is over. Drovers and buyers are staying at the Lochmaddy Hotel. I make supper in my campervan overlooking the pier, turn on the heater, watch the evening light. Rain comes and goes, a northerly breeze stiffens. A fishing boat lands boxes of prawns to a waiting white van at the pier. The *Hebrides* arrives back in the almost-dark after her last sailing of the day. She settles into her berth, bright lights and the purr of overnight generators. An empty bottle is bobbing about at the stern of the prawn boat, up to its neck in water. Cattle bawl intermittently. I have come as close as I will ever get to the island sales. So much has changed since Ian and Neil and Simon were here and yet, today, the crofters brought their animals to market, the buyers came and the drovers and the auctioneers – it's the same, only different. There'll be an early start in the morning. I set the alarm for five o'clock and wriggle into my sleeping bag.

An engine rumbles, someone's shouting, a cow bellows, the alarm clock joins in. They're loading! I dress, drink tea and gobble a marmalade sandwich all at once. Camera, notebook, pen, jacket, hat and out. It's dry with a cold breeze. Dawn is breaking over Skye. The ferry continues to hum. The fishing boat heads out into the loch and I head for the cattle pens.

Two livestock lorries, five men, Kenny with the list of which cattle are going where. He points out that the loading ramp slopes down towards the pens, which means the cattle are climbing uphill into the lorry, especially to the upper decks. The big old cows can be thrawn, he says and just refuse to budge. It's not easy to shift a full-grown cow that's decided not to go anywhere. I remember George Tait saying that, once the last beast was sold, that was it – all hands to work and everyone was a drover. Dingwall workers are joking with the local boys as they bring the cattle forward from the pens. Well wrapped up against the morning cold, the young auctioneer-in-training bears the brunt of the older lads' wisecracks. I think this may be his first trip out to the islands. Despite his white, tired face, he's taking it all in good part.

Kenny's anxious. There should be a third lorry. Where is MacAskill? He should be here by now and he's not answering his phone. Tension rises as the Ahmore lorry fills and the clock ticks towards departure time. Just as the ramp goes up behind the last beast, the missing lorry appears. Ahmore slides into the ferry queue across the road and N. D. MacAskill & Son reverses into the loading bay.

Three huge articulated lorries full of cattle and sheep drive slowly into the mouth of the *Heb*'s vehicle deck with minutes to spare. There's only just enough headroom. My campervan sneaks in behind them. We've made it. I go up to the observation lounge, where the drovers are still hard at work – talking into mobile phones, organising loads, arranging onward transport. Two of the lorries will be going straight to Dingwall when we land at Uig. MacAskill is to deliver a bull somewhere in Skye . . .

'All drivers and their passengers should proceed to the car deck ready for disembarking.'

We're coming into Uig. The drovers are still talking as they go down the stairs: 'What cow does he have? That seven is to go to . . .' It won't stop until they're home tonight. Tomorrow is Sunday. By Monday morning, it'll be business as usual in the civilised warmth of Dingwall's modern mart. I head for my little van crouching in the lee of Ahmore's lorry. The driver is checking anxiously along his vehicle, peering through the ventilation slits into a melee of legs and tails – a drover looking after the beasts in his care.

* * *

I watched as the cattle and sheep of Uist and Benbecula climbed slowly up and round the head of Uig bay. My journey was at an end but the droving went on. I had set out in hope of finding Ian Munro's companions from the 1950s and '60s. I would record their memories, I said, of walking with the cattle, of sales, sea crossings and rail journeys. I would go from one conversation to the next, gathering stories like a drover adding cattle to his herd. Such optimism hid more than a little secret doubt and apprehension but my gatherings had gone deeper and wider than I could ever have predicted. I hadn't planned to go to Crieff or Oban or Lairg. I certainly hadn't planned to walk the North Ford. I had been welcomed and befriended, treated with kindness and generosity. I was bringing home more bright story threads than I had thought possible. It was time to start weaving.

# *Acknowledgements*

*Walking with Cattle* owes its existence to many people. I would like to thank everyone who contributed in any way to my research, which extended well beyond the confines of this book.

Particular thanks are due to Dingwall & Highland Marts Ltd and the Highland Livestock Heritage Society (HLHS), whose Drover Project was the catalyst and whose staff and members – especially Kenneth McKenzie and Janey Clarke – went out of their way to assist my own undertaking.

To all who welcomed me, my notebook and my recorder into their homes and gave so generously of their time, their hospitality and their memories, 'thank you' is inadequate. Every word of every conversation added a vital element to the overall story.

My family and friends believed in the venture throughout and supported me in many and various ways. Thank you all.

Working with the team at Birlinn has been a pleasure. I would like to thank them for their expertise, patience and dedication.

Photographs from the archive of the HLHS are attributed where it has been possible to identify and contact copyright owners. Otherwise, I have credited the HLHS, who have the care of material donated to their Drover Project and held in deposit at the Highland Archive Centre in Inverness.

# *Further Reading*

The following is a small selection from numerous titles that became well-thumbed friends during the preparation of this book. I hope they may prove equally friendly to anyone interested in delving further into the story of droving. Not all are in print but most should be available through the public library service.


Haldane, A.R.B., *The Drove Roads of Scotland*, Thomas Nelson & Sons Ltd, London, 1952, and Birlinn, Edinburgh, 2008.

Keay, J., *Highland Drove*, John Murray, London, 1984.

MacDonald, C., *Life in the Highlands and Islands of Scotland* (combines *Echoes of the Glen*, 1936, and *Highland Journey*, 1943), Mercat Press, Edinburgh, 1999.

MacRae, K.A., *Highland Doorstep*, Moray Press, Edinburgh, 1953.

Moffatt, A., *The Reivers*, Birlinn Ltd, Edinburgh, 2008.

Murray, W.H., *Rob Roy MacGregor: His life and times*, Richard Drew, 1982, and Canongate Press, Edinburgh, 1995.

Stewart, K., *Cattle on a Thousand Hills*, Luath Press, Edinburgh, 2010.

Thomson, J.A., *Ring of Memories*, J. & M. Thomson, Annan, 2004.

Toulson, S., *The Drovers*, Shire, Princes Risborough, 2005.